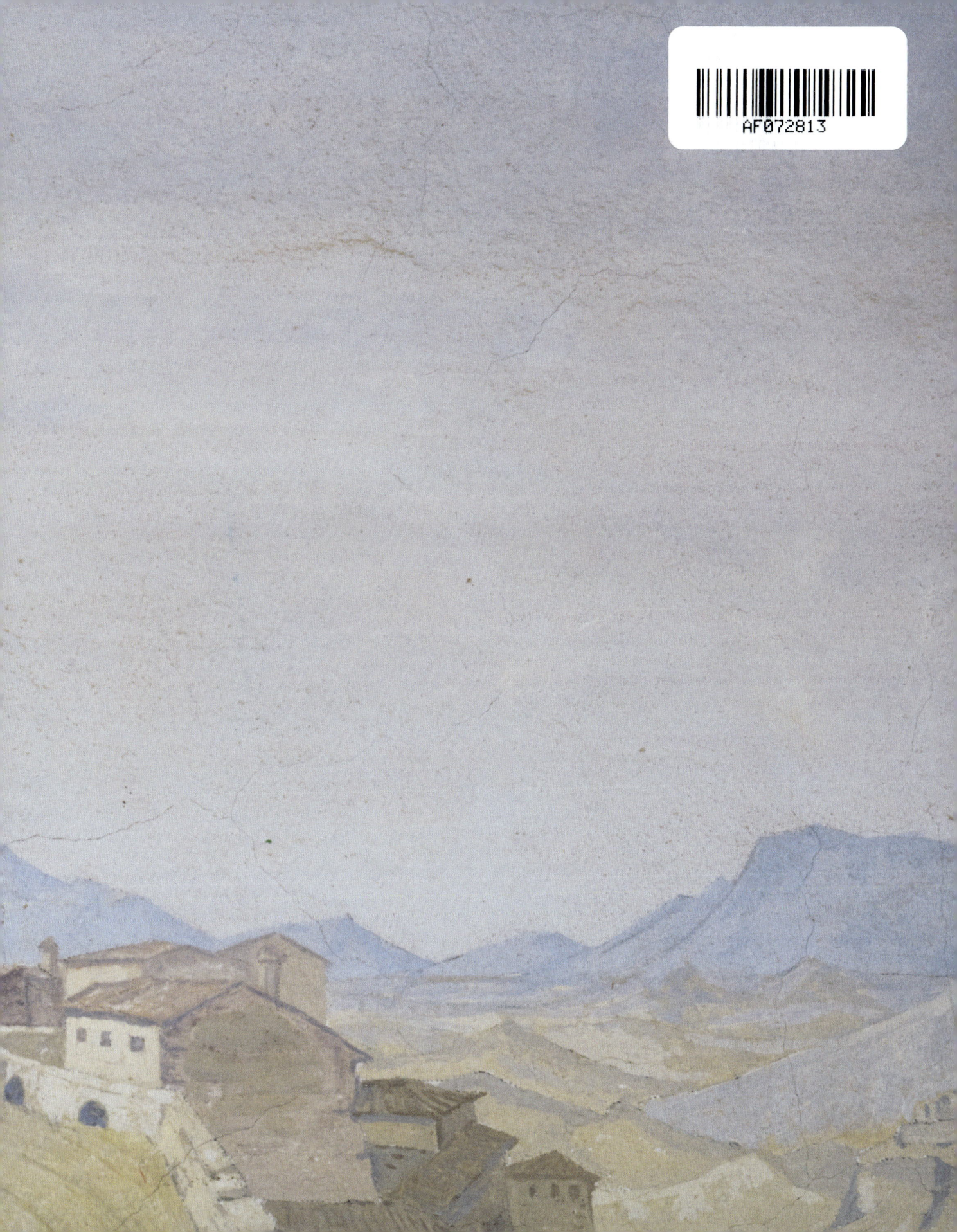

LA FOCE

LA FOCE

PARADISE IN TUSCANY

The Legacy of Iris & Antonio Origo

KATIA LYSY

Prologue by
BENEDETTA ORIGO

Principal photographers
SIMON UPTON *and* MATTEO CARASSALE

This book is dedicated to
ANTONIO LYSY
grandson of Iris and Antonio Origo
1963–2024

CONTENTS

Prologue Memories of a Childhood at La Foce 9
BENEDETTA ORIGO

Map of the property 14

Introduction 19
KATIA LYSY

Chapter 1 Reclaiming the Land 25

Chapter 2 Iris & Antonio 35

Chapter 3 A Pleasant & Well-Proportioned Villa 55

The HOUSE 67
PHOTOGRAPHY BY SIMON UPTON

Chapter 4 A Setting of Beauty, a Labor of Love 141

The GARDEN 161
PHOTOGRAPHY BY MATTEO CARASSALE

Acknowledgments 205

Endnotes 206

Bibliography 209

PROLOGUE

MEMORIES OF A CHILDHOOD AT LA FOCE

"HOW DOES IT FEEL, to live on a Tuscan farm?" Virginia Woolf asked my mother, Iris Origo, who was timidly hoping that her first book would be accepted by the Hogarth Press. "Come and see," was the answer. She did, eventually.

This was in the early 1930s, and the "farm" was La Foce, not yet as we know it now: the garden was still incomplete, the rural tradition was almost medieval, and a teeming agricultural life was very much part of the day. But all this was soon to change, with the war and the social upheaval that succeeded it.

My childhood spanned both worlds. Looking back on my early years, and the privileged way of life that my parents took for granted, I often wonder what they would feel if they could see their (now extremely large) family and La Foce as it has become. I feel lucky to have memories belonging to another century, of a life that in the 1940s and '50s was still the norm—at least in this very conservative part of Tuscany, the Sienese province.

The *fattoria* (central farm) was attached to the main villa, with a courtyard at its core. The olive oil was made here, and so was the wine, which was stored in the cellar under the villa. Farmers came to bring their produce, the pecorino cheese was prepared for winter use, laborers would arrive for their pay, accounts were written out in longhand in great ledgers. Excellent and abundant meals were prepared in the communal *fattoria* kitchen with its enormous oven, where bread and pizza, *la schiacciata*, were made twice a week for quite an army of people, including the refugee children sheltering at La Foce during the war.

La Foce was a completely self-contained little community, producing vegetables and fruit, jam and honey, meat, olive oil, wine, wool, and even saffron from the wild yellow crocus. The *fattoria* was full of people and activity. Quite a crowd would sit down to meals here: the *fattoressa* who cooked and looked after the chickens, rabbits, and geese; several *sottofattori* (the overseer's assistants); the dairy workers and winemaker; gamekeepers and gardeners. On few but exciting occasions, I was allowed to share the delicious *fattoria* lunch, so much tastier than ours, I thought, and the conversation so much more fun.

My father was the ruler here, stern but humorous, active, engaged, admired, obeyed, and respected. He loved this life and the people he worked with, and was dedicated to the reclamation projects he was promoting on his own property and throughout the valley. Wednesdays always put him in a bad mood, because the morning was spent indoors with the accountant. Farming was his passion, and he was completely self-taught—I realized this when, long after he died, I discovered scores of boxes full of agricultural books and farming magazines in an attic.

My mother was very involved with the schoolchildren and their families. Her generosity and kindness became legend. For years, long after her death, on meeting women in the villages I would hear tales of how they had been helped by *la Marchesa*. She was also the best teacher I have ever had. She could transmit her love of reading, poetry, and learning even to a small child, and this pleasure has lasted all my life. Her instructions about writing were to the point: "Only write about something you know. It must have a beginning, a middle, and an end, and you must think before you write."

By the time I was born, in 1940, work on both house and garden was completed, and the main worry was water, always chronically scarce, especially regarding the garden, until my father created a rainwater reservoir to irrigate it. Even so, the grass, the *pratino*, was pretty dry and thirsty by August. Antonio at that point was often heard to exclaim, "We live in the Val d'Orcia. This is not an English lawn!" But, for my younger sister, Donata, and me the garden was always a delight, an inexhaustible supplier of adventure and games: obstacle races in the lower garden and treasure hunts (organized by my mother at Easter), explorations of the woods and the wonderfully wild *crete* (clay hills), making tree houses and playing hare and hounds with the schoolchildren of the *Casa dei Bambini* as our companions. Iris would have her chaise longue rolled out in the summer mornings to the cool wisteria bower, where she would bring a book or her writing materials, and where we would join her for a story or reading aloud. But it is the utter silence of the garden that I remember most, the only sounds in the night being the tinkle of the fountain below my bedroom window, and the occasional hooting of owls from the woods.

The countryside was a never-ending source of wonder. The clay slopes were steep and full of brambles, with wild boar and porcupine paths crisscrossing the way. The heady scent of wild broom was intoxicating, and there were fossils to collect, tunnels made by small animals to peer into, wet clay for making pots to dry in the sun and paint. We would sometimes walk miles to the irrigation reservoirs to swim, and my father would mercifully drive us home for lunch.

In fact, the freedom to roam, both mentally and physically, to invent and discover the world both indoors and out, meant that we were never bored. There was never any sense that danger might lurk around the corner. After nanny age, we were allowed to wander about the countryside, or hide in the attics for hours, and nobody worried—as long as we got home, washed and tidy, in time for meals. Nor were we allowed to be afraid. We girls had to show the courage and strength expected of the boys we should have been—actually a good training for life, and one I have not regretted.

Summer held a special event, the *trebbiatura* (threshing), with our father. The blinding sun seen through a haze of chaff, the noise of the threshing machine, the dust, the oxen, the rides on the hay cart, the cigarettes smoked in secret with the younger farmhands behind the hay stacks, the long tables set out in the shade with white tablecloths, tagliatelle, roast duck and pork, and red wine, and finally the pleasure of returning to our cool home, shutters closed against the sun, are unforgettable memories of quite another era.

Reading was an indoor activity, and my favorite during those long days that in childhood never seem to end. There were plenty of choices in the house, mainly in the small library where books were neatly organized by subject (not language). This was also our living room, where the family would sit by the fire after dinner, each with a book, perhaps watching the news on TV after it was installed in the 1950s, or playing society games such as cards or twenty questions or charades, especially if we had guests, regardless of their age—and grown-ups were expected to participate. Bound collections of English and French Romantic literature were kept in the big salon down a long flight of stairs, sharing the great room with the piano (a Bechstein formerly belonging to my American great-grandmother) and beautiful albums of the great music masterpieces transposed for four-hand piano. What I wonder at today is that I was allowed to choose any book at all, without guidance. If a book would be too difficult for me, I might persevere and come to exciting discoveries, often leading to further exploration.

Another exciting activity was exploring the attics and cupboards with their nooks and treasures. We discovered paintings and discarded drawing albums belonging to our grandfather Clemente Origo, a mysterious artist and draftsman I had never known, as well as dresses and shoes and hats worn by our maternal grandmother and our mother herself, which we used as dressing-up clothes for Christmas plays. At night, however, the house could feel vast and spooky, a feeling that was increased by the ghost stories my mother loved to tell us, while we huddled in bed in excitement and fascination. A much safer place was the *guardaroba* (the ironing room) with its cozy smell of hot damp linen and lavender, and the maids' laughter and gossip. Later, when our winters were spent in Rome so we could go to better schools, the heavy laundry would be brought here to be washed and ironed, and the same truck would return to Rome with delicacies such as wine, oil, pecorino cheese, prosciutto, and firewood.

On Sundays there was Mass, usually in the little cemetery chapel at the end of the woodland path, with the grave of Gianni, our brother who had died of TB at the age of seven, long before I was born. It is a serene and beautiful place, designed by the family friend and architect Cecil Pinsent. Very few people attended Mass there, save a few local families and the schoolchildren. All the men, including my father who towered over everyone else, followed the service absentmindedly from

the outside portico, rocking on their heels, hands behind their backs. When I asked why he wasn't sitting down with us, my father answered that if he didn't come inside he could keep his hat on—as good an answer as any, I thought. After the service all the children would race down the hill and up the next to the La Foce garden and games of hide-and-seek until lunch.

For Christmas, Easter, first communions, weddings, and other major religious holidays and events, services were held in the parish church, a medieval chapel restored and expanded by Pinsent in the ancient Castelluccio, a small castle on the property. On these grand occasions the church would always be full. Music was provided by Iris who pedaled away at the wheezing harmonium behind the altar, accompanying the not-always-tuneful children, including me and Donata, and led by the priest's sister, all the way through the nineteenth-century "Gregorian Mass," which was considered appropriate at the time. Doro, the head *operaio* as well as the husband of my mother's maid, sometimes would join in with his trumpet, as did Dante, the carpenter, on his violin, which

PAGE 8 *Iris Origo loved to write under the small wisteria pergola in the lemon garden.* ABOVE *Origo with her daughter Benedetta, 1941.*

PROLOGUE 11

sounded as though he had made it himself. I always wondered how people could concentrate on their souls during Holy Communion with such a din going on. Our priest was a young man, who gave catechism classes to the schoolchildren and led the boys' *calcio* (soccer) team, too, cassock hauled up to his waist. Sometimes he would invite my father to lunch, protesting all the while that he could offer Antonio only "*un pollaccio cotto nell'acquaccia*" (a meager chicken boiled in plain water). These moments would amuse my father no end.

I was very young in 1943 and '44, when the front passed through our valley, years that were documented in Iris's book *War in Val d'Orcia*. Even so, I have many memories of that time, tinged by childish fears and worries. I remember how my parents would gather in the nursery to listen to the forbidden BBC Radio on a set hidden in the toy cupboard; the arrival from northern Italian towns of frightened and unhappy refugee children; the nights spent cooped up in the cellar under the shelling; my sadness when we discovered that the dog, Gambolino, had been left behind when we were evicted from our house that had been taken over by the fleeing Germans as their headquarters; and especially the "walk" to Montepulciano, refugees in our turn, and being told that I could not hold my mother's hand because I was one of the older children and would have to carry my winter coat, even though it was a hot June day; lying in a ditch while the Allied planes swooped above, strafing the roads; my father, with my baby sister on his shoulders, picking a lucky four-leaf clover and my thinking that we would therefore be safe; my screams at discovering that I had just sat down on an anthill, really the last straw after the long hours of walking and being brave; the cry of "*E' arrivata La Foce!*" (La Foce has arrived!) from the town walls as our friends finally caught sight of our straggling group rounding the hill, and the men running down to meet us and carry the children to safety and shelter. I was not yet four years old.

The End of an Era

The upheaval that took place after the Second World War in Italy, indeed in the world, did not spare La Foce. *Mezzadria* (the sharecropping system that had regulated farming for centuries) came to a painful end. Farmers left their houses and fields to work in small towns, or became small-property owners themselves with government aid. Farm machinery, tractors, and combine harvesters took over all manual work in the fields. Sardinian shepherds bought up land and settled in the valley. Italian landowners were attacked politically and much of their land was expropriated—but fortunately, thanks to my father's historic good management, La Foce was considered a model farm and entirely preserved. Some years later, the European Community became instrumental in providing financial help to agriculture by creating an institution called *agriturismo*, in order to bring life back to abandoned farmhouses and the countryside itself. The valley, under the strong protection of the Val d'Orcia Consortium led by my father, became admired for its uncontaminated beauty. Eventually a natural park was created, and UNESCO claimed it as one of its World Heritage Sites, ensuring, as much as possible, long-lasting protection.

Throughout all these vicissitudes, the Origo family continued to live and work at La Foce. In the 1990s, my sister and I divided up the property (originally of more than seven thousand acres) and sold some of the outlying land. The garden was restored and opened to the public. The villa now has a busy season of weddings and rentals. It is the home of Incontri in Terra di Siena, an internationally acclaimed chamber music festival founded thirty-five years ago by my son, Antonio Lysy. My eldest daughter, Katia Lysy, manages the estate. Agriculture thrives, and our olive oil is particularly appreciated. Restored farmhouses have returned to life, albeit under very different circumstances.

Above all, I am happy to say that my very large family gathers here often, on many occasions. However far they may live and work, everyone considers La Foce to be their home. And this, I think, is its greatest success.

BENEDETTA ORIGO

OPPOSITE *Benedetta Origo in the frescoed dining room at La Foce.*

LA FOCE

LEGEND

1
Villa

2
Fountain Garden

3
Lemon Garden

4
Wisteria Pergola

5
Cypress Allée

6
Statue Allegory of Autumn

7
Lower Garden

8
Statue Allegory of Summer

9
Limonaia

10
Etruscan Necropolis of Tolle

RIGHT
The reclaimed landscape of the Val d'Orcia is the perfect setting for the La Foce gardens.

OPPOSITE
Detail of the garden from Cecil Pinsent's map of the La Foce estate, dated 1941.

INTRODUCTION

KATIA LYSY

It all began in 1924, when my grandparents purchased La Foce. On a blustery October day, Iris and Antonio Origo fell in love with the Val d'Orcia, a spectacularly wild and desolate valley in southern Tuscany, and with La Foce itself, a large estate with a half-ruined fifteenth-century villa at its heart. They dreamed of turning the landscape of arid clay hills, one of the most impoverished areas of Italy, into fertile pastures, wheat fields, olive groves, and vineyards.

As their land reclamation and social projects advanced, so did their restoration of the Renaissance villa and the creation of a new garden, though restoring the house had to take second place to farming and could only advance step by step as agricultural prosperity took hold.

Looking back on the headlong pace of events at La Foce between 1924 and 1940, there is much cause for marvel. So much was done in so little time, and with such an underlying sense of urgency—almost as if Antonio and Iris had a premonition of what was to come, as if they felt the need to seize a golden moment of sentimental, historical, and economic confluence

20   LA FOCE

that would never occur again. A young couple, deeply in love and fired by a profound impulse to make a difference to others, fleeing from the precious aesthetic of their respective upbringings, decided to take advantage of her inherited wealth and his aristocratic authority at a time when a despotic ruler was handing out incentives and prizes for the recovery and enhancement of Italy's rural economy—especially in underprivileged areas like Val d'Orcia—for his own political ends. When war arrived in 1940 and all reclamation work had to stop, Iris and Antonio were ready. Under their leadership, the people of La Foce no longer thought of themselves only as families living in isolated farms under the guidance of a patriarch or *capoccia*; in just fifteen years, they had become a community, headed by the Origos, in which each individual could count on a network of solidarity and mutual assistance. So strong was this community, and so far-reaching its fame, that when hard times came and the winds of war swept through the Val d'Orcia, La Foce was able to offer help and solidarity well beyond its territorial boundaries to the "unending stream of human suffering"[1] that came knocking on its doors.

Over the ensuing century, La Foce has survived private tragedies (the death of the Origos' seven-year-old son Gianni in 1933), public catastrophes (the predations of the German troops in 1943), and epoch-changing events such as the end of the feudal-style sharecropping system, the *mezzadria*, which coincided with Italy's economic boom in the 1960s. A new generation of farmers abandoned the countryside in droves, seeking prosperity in offices and factories. All over Italy, crumbling farmhouses stood abandoned until the 1970s and '80s, when the European Community became instrumental in providing financial help to agriculture by creating an institution called *agriturismo*. The world suddenly discovered an intact rural paradise in the heart of Italy, the Val d'Orcia, and the place where it all began—La Foce. ⊛

PRECEDING PAGES *The clay hills* (crete) *of the Val d'Orcia, 1920s.*
RIGHT *A derelict farmhouse, Val d'Orcia, 1920s.*

CHAPTER 1

RECLAIMING THE LAND

The Val d'Orcia in the 1920s was an expanse of bare clay ridges, treeless and shrubless, running down the valley like a vast lunar landscape, inhuman and wild. The occasional huge boulder interrupted the cracked and rough expanse of pale gray hillocks—the *crete senesi*, "as bare and colourless as elephants' backs, as mountains of the moon."[2] In the background loomed the monumental bulk of Mount Amiata, like some ancient presiding god.

The dilapidated estate of La Foce—then 3,500 acres of shrub oak, eroded clay hills, and twenty-five half-ruined farmhouses set in that bleak, windswept valley—exercised an irresistible attraction on Iris and Antonio Origo, a cosmopolitan young couple in search of their life's work. Friends and family on both sides were appalled and tried to discourage them from buying the property, while neighboring landowners expounded on the heartbreak and expense involved in farming such barren land. Yet after a very few days of riding around the property with the agent-manager, their minds were made up. In her autobiography, Iris recalls:

> *We knew at once that this vast, lonely, uncompromising landscape fascinated and compelled us. To live in the shadow of that mysterious mountain, to arrest the erosion of those steep ridges, to turn this bare clay into wheat-fields, to rebuild these farms and see prosperity return to their inhabitants, to restore the greenness of those mutilated woods—that, we were sure, was the life that we wanted.*[3]

While recognizing the seemingly unsurmountable challenges they faced, Iris wrote to a friend, "It is quite the most beautiful and the wildest bit of country I have ever seen . . . and the whole quality of the beauty is one of loveliness and desolation."[4] In complete agreement, she and Antonio quickly completed the sale and moved to La Foce after their marriage in 1924.

The first years exacted very hard adjustments. Iris did not regret her cosseted but lonely childhood years in the ivory tower that was Villa Medici in Florence, where she was brought up from the age of seven, but she—even more than Antonio—found herself leading a very different life. The leap from the rarefied atmosphere of her mother's intellectual salon, where Edith Wharton, Henry James, Aldous Huxley, and Somerset Maugham were among the illustrious regulars, to the rigors of the Val d'Orcia defies the imagination. No wonder that she confesses to moments of profound discouragement:

> Suddenly an overwhelming wave of longing came over me for the gentle, trim Florentine landscape of my childhood or for green English fields and big trees— and most of all, for a pretty house and garden to come home to in the evening. I felt the landscape around me to be alien, inhuman—built on a scale fit for demi-gods and giants, but not for us. How could we ever succeed

PAGE 24 *The clay hills are now protected and fall within the Val d'Orcia Natural Park, a* UNESCO *World Heritage Site.* ABOVE *In the 1920s and '30s a concerted effort—led by Antonio Origo—was made to turn the arid, stony crete of the Val d'Orcia into farmland.* OPPOSITE TOP *A livestock judging contest, circa 1950.* OPPOSITE BOTTOM *On their first wedding anniversary, in 1925, Iris presented her husband with a pair of young oxen (they turned out to be a very bad buy and had to be resold).*

in taming it, I asked myself, and bring fertility to this desert? Would our life go by in a struggle against insuperable odds?[5]

Despite being an heiress, Iris did not have free access to the fortune inherited from her American father's family, and Antonio's capital had been spent on his share of the purchase of La Foce. Iris reports that there never seemed to be enough money to carry out all the projects requiring urgent attention—even though Iris and Antonio had established, from the very beginning, the priority of farming over all else, as she clearly states in her autobiography:

We both agreed that any plans for the house and garden must give way, for the present, to the needs of the land and the tenants. Anything that the crops

brought in, as well as any gifts from relations, went straight into the land, and I remember that my present to Antonio, on the first anniversary of our wedding-day, was a pair of young oxen which were led under his window, adorned with gilded horns and with silver stars pasted on their flanks.[6]

She continues, in a typically wry tone, "It is sad to have to add that they were such a bad buy that they had to be sold again as soon as possible."

In a paper dated January 1937 and presented to the Florentine agricultural association *I Georgofili*, Antonio explains the program carried out at La Foce by the Origos in their first twelve years:

1) *set up, on each farm, an eight-year rotation*
2) *drain and build sustaining dams on the clay hills to prevent erosion*
3) *increase the arable land*
4) *rebuild the existing farms and the fattoria buildings and annexes*
5) *plant grapevines and olive trees*
6) *build new roads*
7) *build new farms*
8) *increase the livestock and create more pastures*
9) *suspend the cutting down of trees*
10) *increase facilities for education and medical care*[7]

The scope and audacity of their vision is scarcely conceivable today—yet the Origos persisted, overcoming all sorts of obstacles, including their own ignorance and the diffidence of the *contadini* themselves, used to absentee landlords and high-handed *fattori* (overseers).

At the beginning, Iris and Antonio were alone in their efforts, but very substantial help came just a few years later in the form of government subsidies and loans. In 1927 an association of local landowners was founded with Antonio as president, an office he held for over forty years. The *Consorzio per la bonifica della Val d'Orcia* covered almost the entire valley and was entitled to funding offered by the Fascist

government and promoted, in particular, by Mussolini's undersecretary of state for agriculture, Arrigo Serpieri. A passionate advocate of agricultural reform, Serpieri was a great supporter of the concept of *bonifica integrale* (a term that covered all aspects of rural improvement, since land reclamation and water management went hand in hand with hygienic, demographic, social, and economic improvement) and an admirer of the work carried out at La Foce, much of it experimental. In the early 1930s a very wealthy and eccentric American cousin of Iris's died and left her a great deal of money, part of which was used to buy neighboring Castelluccio and its farms in 1934. At this point La Foce was the largest estate in the region south of Siena, counting fifty-seven farms in an area of eight thousand acres.

My grandfather Antonio was very definitely the prime mover in all agricultural decisions, at La Foce and within the *Consorzio*, but Iris was in charge of all matters concerning social, educational, and medical facilities.

A great deal of work went into pulling down and rebuilding the farms, most of which were in a pitiful state.

In the half-ruined farms the roofs leaked, the stairs were worn away, many windows were boarded up or stuffed with rags, and the poverty-stricken families (often consisting of more than twenty souls) were huddled together in dark, airless little rooms. In one of these, a few months later, we found, in the same bed, an old man dying and a woman giving birth to a child.[8]

Schools and kindergartens were built all over the valley, both much needed since on my grandparents' arrival more than 80 percent of the population was illiterate. At La Foce the primary

---

OPPOSITE, TOP TO BOTTOM *Clearing the land of stones, 1930s; digging wells, 1950s; and reforestation, late 1940s.* ABOVE *The first harvest at La Foce, 1925.*

RECLAIMING THE LAND   29

school and the *Casa dei Bambini* (the nursery school, based on the Montessori teaching method) were built close to the main house, but along the road. The children were brought to lessons either in a covered carriage (an unheard-of luxury) with *Casa dei Bambini* emblazoned on the side or by oxcart, when they came from the more remote farms with unpassable roads. After a cooked lunch came a nap on camp beds provided by Iris, and before leaving to go home they were given *merenda*, a substantial snack of bread and butter with prosciutto, brought down from the villa in large wicker baskets.

ABOVE *A horse-drawn carriage was used as a school bus starting in 1934.* OPPOSITE TOP *The classroom in the* Casa dei Bambini *nursery school at La Foce.* OPPOSITE BOTTOM *Each day Iris provided the schoolchildren with a hot meal, followed by a nap on camp beds, 1930s.*

By the early 1930s the school at La Foce had seventy-five pupils whose curriculum reflected Iris's views on education rather than those of the Fascist government. In fact, apart from the obligatory portraits of the king and of Mussolini hanging in the classroom and the occasional parade on feast days in military-style uniforms, the children's education was singularly lacking in Fascist elements. The teachers were told never to shout at the children, to read to them as much as possible, and to encourage them to spend a lot of time outdoors. Each school had its own experimental field and garden where the children could prepare for their future lives as farmers, but also grow to love plants and flowers.

> The children's pride in their new schoolrooms was delightful to witness. I remember that, when the

30   LA FOCE

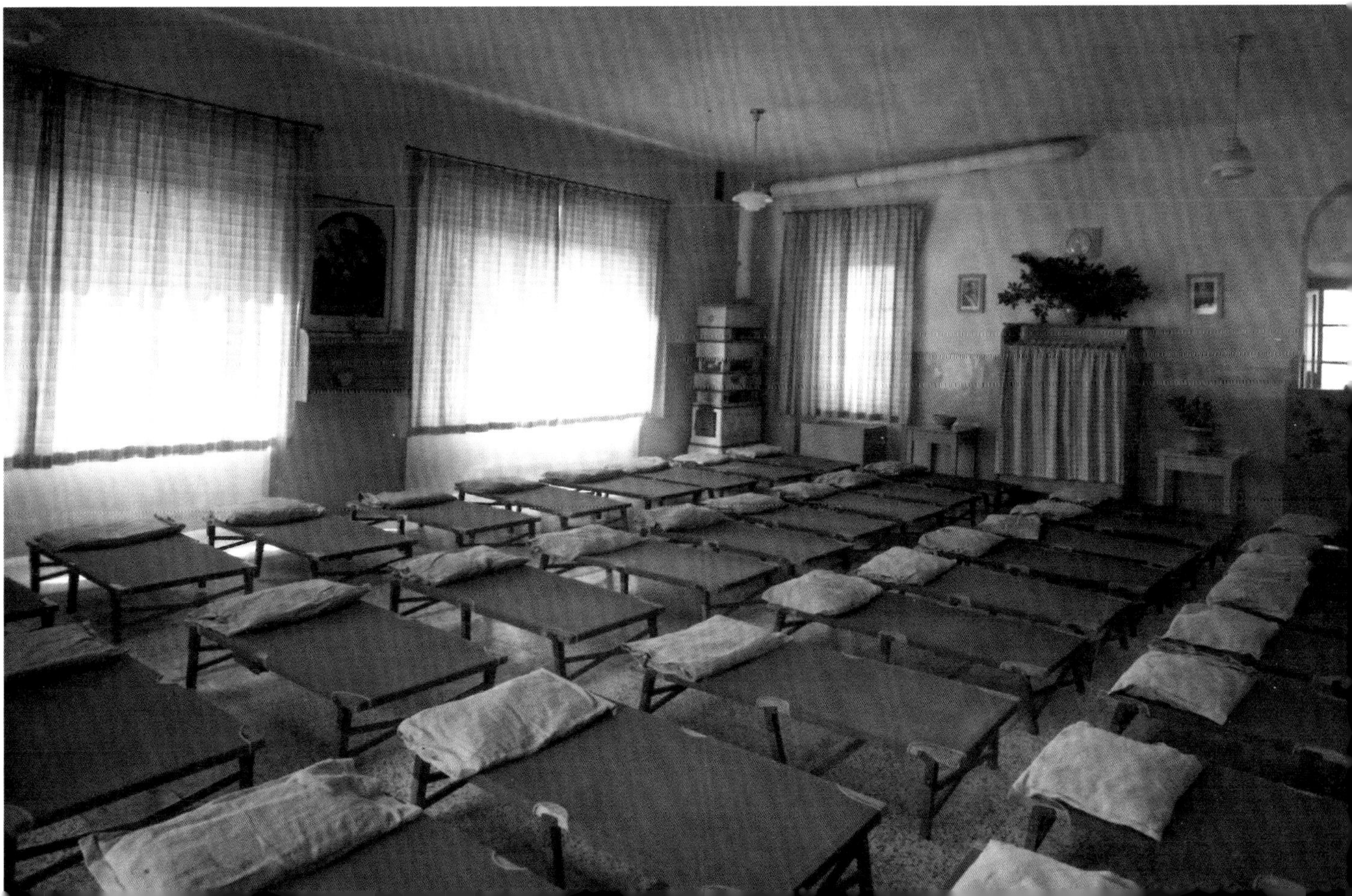

one at La Foce was opened (its walls painted in gay colors and adorned with pictures and maps) we found the pupils, of their own accord, taking off their muddy boots before coming in, so as not to sully the shining floors.[9]

In January 1943 the *Casa dei Bambini* became a shelter for twenty-three refugee children from Turin and Genoa whose homes had been destroyed by bombs. Iris tells their story, and that of La Foce in the war years of 1943 and 1944, in her best-selling diary, *War in Val d'Orcia*. Later on, in the 1950s and '60s, the *Casa dei Bambini* became a foster home for a small group of orphans who called Iris "Nonna."

Almost as much as schools, the Val d'Orcia needed hospitals and medical care. In 1933 a small clinic was built at La Foce in memory of the Origos' son Gianni, who died of tubercular meningitis shortly before his eighth birthday. A resident nurse was installed in the clinic, and she traveled all over the estate, vaccinating the children and visiting even the most remote farms. There were beds for emergencies and childbirth, and the very latest ultraviolet-ray lamps for curing rickets, which was common among the undernourished children of the Val d'Orcia.

Leisure time was also considered important as a way of bringing the community together. In 1939 *Dopolavoro La Foce* was built, providing a recreational club with a small shop where La Foce workmen could meet for a glass of wine and a game of bocce in the shade of the lime trees. Iris loved putting on plays, and a small stage painted with figures of Harlequin and Columbine was set up on special occasions for the schoolchildren to perform on. All of the costumes and scenery were made by the La Foce mothers and carpenters—I still have a giant blue teapot that once featured in *Alice in Wonderland* and half a broken wooden donkey used in a Nativity play.

When Italy joined the war in 1940 all work was suspended, including on the house and garden.

During the war our little community, always largely self-contained, became almost entirely so, held together by a bond of common interests, anxieties, fears and hopes. Together we planned how to hide the oil, the hams and the cheeses, so that the Germans would not find them; together we found shelter and clothes for the fugitives who knocked at our door—whether Italian or Allies, soldiers or civilians—together we watched the first bombs fall on

the Val d'Orcia bridges, and listened hopefully to the rumours of landings in Tuscany, which never came. And together, when the Germans had turned us out, we returned—after the Allies' arrival—to reap the harvest, to start clearing the mines and rebuilding the shattered farms.[10]

On June 22, 1944, Iris and Antonio were forced to flee their home in a desperate attempt to bring the twenty-three refugee children in their care to safety. In War in Val d'Orcia, Iris ends on a moving note of encouragement and hope.

The day will come when at last the boys will return to their ploughs, and the dusty clay-hills of the Val d'Orcia will again "blossom like the rose." Destruction and death have visited us, but now—there is hope in the air.[11]

Under the guidance of Iris and Antonio, the bleak, stony clay hills of the Val d'Orcia—swept by the north tramontana wind in winter and enveloped in swirling dust in summer, with nothing but mud and thick bushes of broom in the seasons in between—blossomed into a full-blown pastoral utopia. Today's visitors to the Val d'Orcia come for the beautiful landscape, seen as perfectly reflecting the Renaissance ideal of a harmonious coexistence between man and nature. The truth, as always, is more complicated: the desperate poverty of this land paradoxically turned out to be its salvation, providing a tabula rasa for Iris and Antonio's enlightened development. ⊛

OPPOSITE LEFT *In the 1950s and '60s, the* Casa dei Bambini *became a foster home for a small group of orphans, who called Iris* Nonna. OPPOSITE RIGHT *Iris (back row, center) with refugee children in the winter of 1943–44.* ABOVE *A small stage painted with the figures of Harlequin and Columbine was set up at the* Dopolavoro La Foce *for the schoolchildren to perform on.*

CHAPTER 2

IRIS & ANTONIO

Iris Cutting was born in England in 1902 to an Anglo-Irish aristocrat, Lady Sybil Cuffe, and an American father, William Bayard Cutting Jr., descended from one of New York's oldest and wealthiest families. My grandmother's life was one of great privilege—"I was tipped with fairy gold"[12] she said of herself—and of these "unfair advantages of birth, education, money, environment and opportunity"[13] she was very conscious. Hers was also a life of great intensity and variety, of worldwide travel and beautiful homes and—at least twice—of major tragedy.

When my great-grandfather Bayard was twenty-two and newly married, he contracted tuberculosis, which struck three of the four Cutting siblings. (Bronson and Olivia eventually recovered.) Consequently, Iris's first years of life were spent traveling all over the world with her parents in search of a miraculous cure. From California to Saint Moritz, from the Adirondacks to Portofino, the young family migrated year round from hotels to chalets to sanatoriums. All to no avail—Bayard died in December 1909 along the Nile River in a luxuriously appointed *dahabeah*, a present from his father, and was buried in Aswan. Bayard was only thirty, and Iris, his beloved only child, was seven. A brilliant man and learned scholar who served as a diplomat to the Court of St. James's, he excelled in numerous fields and was

PAGE 34
The library at La Foce.

LEFT
Lady Sybil Cuffe, Iris's mother, was an Anglo-Irish aristocrat, 1898.

OPPOSITE
William Bayard Cutting Jr., Iris's American father, in 1904 when he was thirty-one years old, six years before he died of tuberculosis.

I remember the spring day on which my mother took me for a drive up a long hill, first between high walls over which yellow banksia roses tumbled and a tangle of wisteria, then through olive-groves opening to an ever wider view; and finally down a long drive over-shadowed by ilex trees to a terrace with two tall trees—paulownias— which had scattered on the lawn mauve flowers I had never seen before. At the end of the terrace stood a square house with a deep loggia, looking due west towards the sunset over the whole valley of the Arno.[15]

In the eighteenth century, the Villa Medici was owned by Horace Walpole's sister-in-law, who covered three rooms in exquisite Chinese wallpaper of flowers and birds in brilliant colors. When Sybil moved in, she asked her friends Geoffrey Scott and Cecil Pinsent (then engaged in a lengthy and tormented restoration of the neighboring Villa I Tatti, where the famous art historian Bernard Berenson, or BB as he was known, and his wife, Mary, lived) to help her furnish the house and restore the original plan of the garden. In point of fact it was Pinsent rather than Scott who helped Sybil with the decor, designing a beautiful dark-red and gold Chinese-style library for her, bringing tiles from Vietri in the green and sea-blue colors she loved, introducing her to the delights of shopping for the Renaissance chests of drawers, rococo gold mirrors, and intricately patterned brocades eagerly displayed by Florentine noble families in need of cash.

Villa Medici was soon exquisitely appointed and Sybil— elegant, fragile, and capricious—became a much-admired hostess. A constant stream of visitors, especially from the Anglo–Florentine community, came to gossip, comment, and play elaborate drawing-room games mainly aimed at showing off one's cultural knowledge. Henry James, Edith Wharton, Aldous Huxley, Vernon Lee, and members of the Bloomsbury group visited, and poor little Iris, who was kept at home with a string of governesses despite her burning desire to go to school

missed by many, from the philosopher George Santayana, whose pupil he was at Harvard, to Edith Wharton, a great friend despite the difference in their ages, who wrote a touching, privately printed piece in his memory.

As he lay dying, Bayard composed a letter to Sybil that was to have a lasting influence on Iris's life.

All this national feeling makes people so unhappy. Bring [Iris] up somewhere where she does not belong, then she can't have it. I'd rather France or Italy than England, so that she should really be cosmopolitan, from deep down.[14]

Bayard's pretty, hypochondriac widow respected his dying wishes, and despite quite a few transatlantic tussles with the New York side of the family, Iris ended up living far from both England and America.

In 1911 Sybil bought the beautiful Villa Medici in the hills above Florence, where the flower of Tuscan humanists gathered around Lorenzo the Magnificent. The villa's natural surroundings made a deep and lasting impression on the young Iris.

ABOVE *Bayard Cutting with two-year-old Iris, 1904.* OPPOSITE *One of the loggias of Villa Medici in Fiesole, near Florence, adorned with Renaissance furniture and terra-cotta pots planted with* Dahlia imperialis.

PRECEDING PAGES
Villa Medici in Fiesole, with the lower garden (right), redesigned by Cecil Pinsent for Sybil Cutting, who purchased the villa in 1911.

LEFT
The sitting room at Villa Medici.

and make some friends, took the guests on historical tours of the house. In short, Iris perceived Villa Medici as a beautiful but suffocating ivory tower, totally estranged from the political reality of the country they lived in, from where the First World War was perceived only as "an unpleasant noise off stage."[16]

Villa Medici was one of three fixed points in Iris's childhood, each one very different from the other, except on the scale of privilege and luxury.

Westbrook, the sprawling Tudor revival house on Long Island built for Iris's grandfather William Bayard Cutting Sr., was an earthly paradise with infinite delights for a child.

There was so much to do and see: the thickly-wooded islands to explore . . . the river for canoeing, the barns on the home-farm which housed the fine herd of Jersey cows, the long-legged nuzzling calves, the trees of the arboretum, some of them with low, spreading branches which formed a green tent into which one could creep and lie, savouring the aromatic secrecy and darkness.[17]

Yet Westbrook was also a very formal place, with perfectly trimmed shrubs, croquet lawns, and legions of servants—one of the great houses of Old New York, straight from the pages of *The Age of Innocence*. As time went by, the world changed but Westbrook remained the same, so much so that Iris commented during a visit with her daughters:

It was impossible not to feel—especially coming from war-time Europe, where so many great houses had

met with destruction, and so many others, though still standing, could not hope to return to the life of the past—that we were existing in a world without a future, one which only my grandmother's presence rendered justifiable at all.[18]

Yet Iris later acknowledged that she was wrong—Westbrook proved to have a future after all. With their characteristic combination of liberality and good sense, the Cuttings donated the house and the grounds, along with a generous endowment, to the state of New York for use as a public botanical park and garden. (It is now the Bayard Cutting Arboretum.) Iris's Cutting grandparents, both devout Episcopalians, had a strong sense of charitable duty and devoted a great deal of their time and money to helping associations and individuals in need—a lesson Iris learned from a very early age. (When she was just nine, in the winter of 1911, her American grandfather wrote to tell her she would not be receiving a Christmas present from him that year; instead, he planned to give a present and Christmas dinner to every child in one or two orphanages in New York. "It will be for them the 'Iris Cutting' Christmas,"[19] he wrote.) Many decades later in Val d'Orcia, the Cutting lesson bore fruit.

The other point of reference for the young Iris was Desart Court, the Irish home of her maternal grandfather Hamilton

OPPOSITE *The Cutting family at Westbrook, now the Bayard Cutting Arboretum, Great River, New York, circa 1900.* ABOVE *Desart Court, County Kilkenny, Ireland, home of Iris's maternal grandparents, burned down during the Troubles in 1922.*

IRIS & ANTONIO

LEFT
Lord and Lady Desart at Desart Court, 1904.

OPPOSITE,
TOP TO BOTTOM
Antonio Origo as a young man, circa 1915; supervising the olive oil production in the 1950s; and in the garden at La Foce in the 1970s.

Cuffe, Lord Desart, affectionately known to her as "Gabba." Desart Court was a place of happiness. Playing with her Verney cousins, the children of Sybil's sister Joan and brother-in-law Harry Lloyd Verney, was a highlight of Iris's lonely life as the only child of a self-centered, if fascinating, mother, and the long summer holidays she spent at Desart Court were filled with riding and walks and picnics:

> *It was the freedom that was the real delight; to explore the woody tangle of laburnum and laurel in the shrubbery, where we had a "secret" hut, thatched with branches and carpeted with hay; to climb the apple-trees and pear-trees and munch their sunny fruit, to bicycle, when I was a little older, on the road outside the lodge gates. . . .*[20] *So each summer passed—serene, immutable, unending. Were they really so many or so long? . . . It hardly matters how many they were. Green sleeping parkland, deep woods, peaches on a sunlit wall, laughter and freedom—they have been enough to fill a lifetime.*[21]

As a child, Iris could not perceive the growing unrest and indeed hatred simmering in Anglo–Irish politics, just as she was unaware of Lord Desart's growing desperation as he came to realize the impossibility of conciliating his profound love for both Ireland and England: the shadow of "the Troubles" had already fallen and was to culminate in the destruction of Desart Court, which burned down in 1922. "The wound is deep and there is no cure for it,"[22] Desart wrote to Iris. He never returned to Ireland.

Three houses and gardens—Westbrook, Desart Court, and Villa Medici—all held central roles in the life of the young Iris, but Villa Medici was destined to exercise the greatest influence in the years to come. It was there that Iris met Pinsent, whom she turned to many years later when creating a garden of her own at La Foce.

Many readers of Iris Origo have commented on the fact that her husband is a shadowy, little-explored character in her books. It is true that she rarely mentions my grandfather Antonio—the imposing yet affectionate figure I remember from childhood who is so frequently evoked by his daughters in family anecdotes and memories. Iris's biographer, Caroline Moorehead, ran into the same difficulty and is regretfully reduced to commenting, "Though he was obviously a man of great charm and warmth, intelligent and fond of his friends, it's often hard to get any real sense of him."[23]

In fact, Antonio was the cornerstone without which the entire construction that was the estate of La Foce—agriculture and social—would have come tumbling down or, indeed, could never have been erected. My grandmother's philanthropic projects would never have taken off without his pragmatic, far-seeing approach to the many obstacles that constantly arose.

Antonio was many people at once: the illegitimate son of the Marchese Origo, bundled out of the way for much of his childhood until his parents were able to marry and his claim to the title was recognized; a debonair aristocrat much admired in the salons of the Roman nobility; a suave diplomat who had the ear of Italy's minister for agriculture (though his permit to import an American-produced tractor for La Foce was revoked in 1939); a passionate farmer who experimented with new techniques and spurred the other, largely conservative landowners in the province of Siena to innovations (though not all of his projects were equally successful—the Simmental cows needed greener, fresher pastures than the Val d'Orcia could provide and the Angus sheep brought over from Scotland to be crossed with the local breed succumbed to tick fever). Above all, a disciplined, level-headed decision-maker, the person everyone turned to in wartime. He could meet the German officers on their own ground thanks to his military career and excellent command of the German language; reassure the anguished peasants who turned to him for advice and help; meet with the local partisans and their leader, Beppe; and stop them from making grand but dangerous gestures bound to provoke German reprisals. He was able to convince the local authorities and especially the *Carabinieri*, whose lack of enthusiasm for the Nazi invaders was clear, to entrust him with a group of British prisoners of war, whom he agreed to keep under his supervision (only to organize their escape on arrival of the Germans). My grandmother—due

to her gender and even more so to her nationality—could never have acted as boldly, and her help was mostly carried out behind the scenes. When the official Fascist newspaper of Siena published a virulent article against Iris (describing her as "a very rich American–English woman who does as she pleases in her domain and does not heed the authorities"),[24] Antonio tackled first the Prefect of Siena and then the colonel of the Fascist militia. However obliquely described, Antonio, with his understated bravery and good sense, is always present in Iris's *War in Val d'Orcia*.

Both Iris and Antonio believed deeply in respecting each other's privacy. Antonio was immensely proud of his wife's talent but would never have sanctioned speculation or revelations of any sort about his beliefs or activities—not regarding his political sympathies, his duties as president of the *Consorzio della Bonifica*, his growing sense of shame and betrayal as the king and the Fascist government led Italy into the disaster of war, or his role as champion of the farmers at La Foce and protector of escaped prisoners of war, partisans, and refugees. Like his father before him, he was a cavalry officer with the *Genova Cavalleria*; Antonio was awarded a bronze medal for valor in the grueling trenches of the Karst Plateau during the First World War. A natural leader, he possessed an authority and personal charisma that stood him in good stead whether he was negotiating with government authorities as head of the *bonifica* project in the Val d'Orcia or dealing with the German invaders. When the Allies found themselves facing the gargantuan task of reordering

ABOVE *Iris and Antonio departing from Villa Medici on their wedding day, 1924.* OPPOSITE *Iris and her son, Gianni, who died in 1933 before he was eight.*

the shambles that was postwar Italy, in the Val d'Orcia they turned to Antonio, the only person who commanded the respect of the many competing Italian political factions and the trust of the Allied leaders. Out of the sense of patriotism and duty that was such an important part of his character, Antonio accepted the post of mayor of Chianciano for the period immediately after the war, until local elections could be organized. From what I remember of him and from his daughters' stories, it must have been an immense sacrifice for him to take time from his beloved land in order to act as peacemaker in the countless petty local quarrels, such as those among the hotel owners and shopkeepers who were struggling to regain Chianciano's prewar prosperity as a well-known spa resort.

A man of action, an admirer of the arts, and fluent in four languages, Antonio was brought up in a cultivated, aristocratic household. His Russian-born mother was a singer from a noble Milanese family, and his father, the Marchese Clemente Origo, a well-known sculptor and painter. Their house was a meeting place for famous musicians, painters, and poets, among them Giacomo Puccini and Gabriele D'Annunzio. Iris, a true intellectual, was undoubtedly fascinated by Antonio's mixture of culture and pragmatism and by the prospect of spending her life with a partner with whom she could put her philanthropic ideals into practice.

When Antonio and Iris met, she was eighteen and he was twenty-nine. After a long engagement and much opposition from Iris's mother (supposedly because of the differences in background, religion, and age, but actually because Sybil was going through a difficult divorce from her second husband, Geoffrey Scott, and was afraid of being left alone), they were married at Villa Medici in March 1924. Antonio's relatives were among the many guests—most of whom, Iris wrote, "will belong to that section of the British colony which looks as if it had been buried and dug up again for the occasion"[25]—but not Sybil, who took to her bed with a real or imaginary illness.

Iris and Antonio spent long periods at La Foce, but they loved traveling and often went on adventurous expeditions all over the world or paid visits to faraway friends and relatives. In

1925 the Origos had a son, Gian Clemente, who was known as Gianni. He loved spending time at La Foce, where he had his own little garden full of flowers, a pony, and dogs; both parents found it increasingly difficult to embark on their customary travels and leave Gianni with his nanny.

In 1933 disaster struck. Gianni fell ill with tubercular meningitis and died—by a tragic coincidence he was the same age that Iris had been when her father died. The greatest sorrow of her childhood was repeated and magnified a hundredfold. Iris and Antonio mourned their little boy in different ways; Antonio threw himself into farming work, while Iris wrote grief-stricken letters to friends:

Everything is so lovely here now—and the loveliness is almost unbearable. It is in the simplest things—the wind in one's face, the smell of the pinks in the

garden—that it seems most intolerable, most against nature, that he should never feel again.[26]

In the terrible months following Gianni's death, Iris tried to find consolation in the company of dear friends, in planning the garden, and in putting together a heartbreaking book about her son, in which she recorded every detail of his life. Privately printed on thick cream paper and beautifully bound, it is full of photographs separated by sheets of white tissue. She sent copies only to family members and a small circle of friends. Many years later, on coming back home to La Foce to view the damage caused by the retreating Germans, she was deeply moved to find her own copy of the book waiting for her.

We had expected, when we left, to find our house destroyed by shell-fire, but instead found no more serious damage than a few large holes in the roof, the cutting-off of water and of light . . . the removal and destruction of any window-pane or door, a few bullet-holes in books and pictures, and an all-pervading stench of excrement and refuse. But on the hall-table, intact but for a few mud-stains, lay a copy of the book which, after the death of our son Gianni, I had written as a private record of his short life, and of which a few copies had been printed for relations and close friends. On the flyleaf a pencilled note told me that this book had been found in the woods (left there by the Moroccan troops with the Fifth Army, the Goums, who had ransacked everything they could lay hands on) by an English soldier, who "realising it must be of great sentimental value" had obtained permission to walk back many miles with it, so that we might find it when we came home. This piece of imaginative kindness, at such a moment, was so consoling as to outweigh every other loss.[27]

Iris also found consolation in the project of building a small cemetery at La Foce, especially for Gianni (who died in Villa Medici and was buried in Fiesole) and the people who lived

RIGHT *The La Foce cemetery.*

and worked on the estate. Pinsent designed a series of terraces and a beautifully simple chapel in the same local travertine used for the La Foce garden, just a few minutes' walk from the house through the woods and with a beautiful view of the Val d'Orcia that is framed by the branches of a great oak tree. Six months after his death, Gianni's coffin was brought home. Iris writes:

> *It was a day of utter beauty, cloudless, no leaf stirring. All our farm-people, and many from farms and villages far across the valley, were waiting with us in front of the house, and with them all the school-children, each one carrying a bunch of flowers. . . . The little coffin was lifted down and carried by the members of the household and the farmers whom he knew best— preceded by some of the workmen with the wreaths, by the priest, and by four little boys, Ugo [Gianni's best friend, the gardener's son] among them, carrying lighted candles. As the coffin passed among the people, everyone knelt—then slowly they formed into procession behind. The school-children were on each side of him, then we followed, then all the farmers— and so we slowly made our way down the road, among the olives and the vines and the freshly-ploughed fields, and through the little oak-wood up the hill, until we came to the chapel. There the prayers were said for him, with sunlight pouring onto the altar. Antonio, Dino, Cecil and the fattore carried him out and laid him in the grave lined with boughs of cypress and of lavender, and the school-children, one by one, threw in their bunches of flowers. He is in his own land now, among his own people. "I want to go back to La Foce because it's so peaceful there."*[28]

Gianni was deeply attached to La Foce, and this was his stock phrase when he was tired of traveling and wanted to go home.

LEFT, TOP TO BOTTOM *Iris with Gianni in the fountain garden, La Foce, 1931; with her friend Elsa Dallolio and daughter Benedetta, 1941; and with her daughters, Donata and Benedetta, at Lerici, 1950.* OPPOSITE *Iris at work, circa 1934.*

In the years immediately after Gianni's death, Iris found some distraction from her grief in reading, visiting friends, and writing books. She spent long periods in England, far removed from the practical agricultural and social problems of the Val d'Orcia. In London she revived old friendships and formed new ones, including with Virginia and Leonard Woolf; attended pacifist rallies; and supported the Kindertransport scheme to bring European Jewish children to safety in Great Britain. In 1935 she wrote the short, tragic life of Lord Byron's little daughter Allegra. In 1935 and 1938 she completed two biographies published by the Hogarth Press, one of Giacomo Leopardi, the tortured nineteenth-century Italian poet and philosopher, and the other of Cola di Rienzo, the medieval revolutionary and demagogue who advocated the return to Rome's imperial greatness and inspired verses by Petrarch and Byron. These books received high praise and marked the start of Iris's literary career as a biographer and historian (*The Last Attachment: The Story of Byron and Countess Guiccioli*, 1949; *The Merchant of Prato*, 1957; *The World of San Bernardino*, 1963; and *A Need to Testify*, 1984; among others), diarist (*War in Val d'Orcia*, 1947, and the posthumously published *A Chill in the Air*, 2017), and memoirist (*Images and Shadows*, 1970).

Yet, despite prolonged sojourns in England, she never abandoned La Foce or her husband. As the political situation worsened and Mussolini's campaign against the "decadent democracies" intensified, Iris realized she must choose between her country of birth and her adoptive home, which included Antonio and La Foce. In 1938 she handed in her American passport—the original document, with "CANCELLED" diagonally stamped across it in large letters, is preserved in the La Foce archives—and cast her lot with her Italian husband and the Italian people.

Seven years after Gianni's death, the Origos had two daughters: Benedetta, born in 1940, and Donata, in 1943. ⊛

CHAPTER 3

A PLEASANT &
WELL-PROPORTIONED VILLA

Even before their marriage, Iris and Antonio were hunting for a country home of their own. It had to be far from their respective families—the Origos in Rome and Sybil and her brilliant, over-sophisticated, backbiting circle of expatriates in Florence. Despite their high-society life among artists and aristocrats, Antonio's parents had chosen a very different path for their son—and one contrary to his true passion, which was for agriculture. They decided he should train as a businessman, and after he finished boarding school in Switzerland they sent him first to work in a bank in Brussels, then for a spell at G. H. Mumm, the Champagne firm in Reims. Right from the start, both Antonio and Iris were quite certain of what they wanted for their shared future: a life entirely new to both, a life with meaning and purpose. Iris is very clear on the subject:

> *Antonio had, deep in his bones, the instinctive love of many Italians for the land, and wanted to farm in a region still undeveloped agriculturally, where there would still be much work to do. I had a strong, though uninformed, interest in social work. We both wanted to get away from city life and to lead what we thought of as a pastoral, Virgilian existence.*[29]

During Iris's childhood in Fiesole, Bernard Berenson would frequently call upon Sybil, accompanied by his many dazzling guests, and take her driving in the hills around Florence or on longer expeditions to discover new sights:

PAGE 54
A view of the house from the lemon garden.

LEFT
The villa in 1925, when work on the facade and the entrance had already begun.

a remote Romanesque church, a little-known fresco, a striking view. Though reluctant to be included in this alarmingly sophisticated grown-up company, Iris usually went along, too. As a result, she was very familiar with much of Tuscany, but even after looking at many estates for sale there, neither she nor Antonio found what they were seeking, "a place with enough work to fill our lifetime . . . in a setting of some beauty."[30] Iris envisioned their home clearly:

> *one of the fourteenth- or fifteenth-century villas which were then almost as much a part of the Tuscan landscape as the hills on which they stood or the long cypress avenues which led up to them: villas with an austere façade broken only by a deep loggia, high vaulted rooms of perfect proportions, great stone fireplaces, perhaps a little courtyard with a well, and a garden with a fountain and an overgrown hedge of box. What I had not realised, until we started our search, was that such places are only likely to be found on land that had already been tilled for centuries, with terraced hillsides planted with olive trees and vineyards that were already fruitful and trim in the days of the Decameron. To choose such an estate would mean that we would only have to follow the course of established custom, handing over all the hard work to our* fattore, *and casting an occasional paternal eye over what was being done, as it always had been done. This was not what we wanted.*[31]

Finally, they happened upon La Foce, a vast estate south of Siena near what was then the new little spa town of Chianciano, in a water-starved, windswept valley of clay hills, stony and bleak yet fascinating in its majestic beauty. The farming land was very poor and the villa itself was pleasant and well proportioned, but not imposing and rather run-down.

Originally built in 1498 by the Hospital of Santa Maria della Scala, the greatest Sienese landowner at the time, as a posthouse or tavern, La Foce is strategically placed at the crossing of three roads frequented by merchants and pilgrims traveling along the nearby Via Francigena and Via Cassia. Centuries before the tavern was built, however, La Foce was the site of a large and prosperous Etruscan farm and part of a larger settlement, as testified by the vast necropolis on the small hill named Tolle facing today's villa. Hundreds of tombs spanning a period from the eighth century BC to the second century AD have been brought to light quite recently

RIGHT
Maremmana cattle in the courtyard of La Foce, before Pinsent's 1931–32 restoration.

by volunteers from the Chianciano Archaeological Society, and the finds are on display in the Museum of Chianciano. In the nineteenth century, excavations carried out illegally throughout the Chiusi-Chianciano area by *tombaroli* (grave robbers) yielded terra-cotta reliefs, bucchero ware, and canopic jars, many of which found their way to private collections in Italy and abroad. Leone Mieli, from whose son Aldo my grandparents purchased the estate in 1924, was a particularly passionate collector—but at least he donated his finds to the city of Siena in 1882.

The Romans took over from the Etruscans, expanding and developing the area's rich agriculture by cultivating spelt and wheat and planting vineyards and olive groves. Even the barbaric invasions and subsequent medieval pillaging at the hands of Lombards, Carolingians, and Normans could not entirely destroy the agricultural riches and dense woods of the Val d'Orcia. Indeed, so prosperous was the land that in the fifteenth and early sixteenth centuries it was known as *il granaio di Siena*, Siena's granary. Scholars have recently confirmed that the clay hills of the Val d'Orcia emerged relatively recently, not more than 350 or 400 years ago, and that the lunar landscape my grandparents found was caused not by climate change, but by the constant wars between Siena and Florence, when pillaging and deforestation drove the Sienese countryside to ruin. Then came the plague of 1630, which left a large part of the Val d'Orcia virtually uninhabited. The sharecroppers, or *mezzadri*, sank deep into debt, and when the Dei family acquired the estate from the hospital in 1736, they were unable to improve the situation, which was exacerbated by a free-trade policy enacted by the dukes of Lorraine that caused the price of wheat to collapse. When the estate went bankrupt in 1837 it was bought by three brothers, bankers of Roman–Jewish origin named Abramo, Leone, and Tranquillo Mieli, who attempted to introduce new farming techniques and cultivations. Eventually La Foce was put on the market again, and finally purchased by the Origos. Iris wrote in 1924:

The home itself was certainly not the beautiful villa I had hoped for, but merely a medium-sized country house of quite pleasant proportions, adorned by a loggia on the ground floor, with arches of red brick and a façade with windows framed in the same material. Indoors it had no especial character or charm. A steep stone staircase led straight into a dark central room, lit only by red and blue panes of Victorian glass inserted

in the doors, and the smaller rooms leading out of it were papered in dingy, faded colours. The doors were of deal or yellow pitch-pine, the floor of unwaxed, half-broken bricks, and there was a general aroma of must, dust and decay. There was no garden, since the well was only sufficient for drinking-water, and of course no bathroom. There was no electric light, central heating or telephone.[32]

What a contrast to today's airy, luminous house, with its frescoes and trompe-l'oeil moldings, terra-cotta floors, and travertine steps and arches. The transformation was wrought by Pinsent, who began work on the house immediately after it was purchased (while my grandparents wisely absented themselves on a long honeymoon). Iris trusted Pinsent, a close family friend since her childhood, implicitly. He and Geoffrey Scott, who was to become Sybil's second husband, had worked together on I Tatti and shared a neo-humanist aesthetic (Scott's *Architecture of Humanism* is still considered a classic), and Pinsent had helped Sybil restore the garden at Villa Medici. What could be more natural for Iris and Antonio than to call on him for help in rebuilding the villa at La Foce and creating a beautiful garden?

The first great drawback at La Foce was the lack of water, and a close second was the challenging geographical location, situated as it was on a dry, windy hilltop that overlooked the *crete senesi* and tilted not just downwards but also sideways. The immediate creation of a garden, though greatly wished for, was therefore abandoned so that Pinsent could concentrate on making the villa livable—and less gloomy. By the time the newlyweds came back from their honeymoon, the main hall had been converted into a dining room with a large skylight, the walls were painted in light colors, a bathroom was installed, and beautiful oak bookcases were made for the library, which doubled as a small sitting room.

In the main entrance hall, the floor was paved with dark terra-cotta tiles outlined in slabs of the local travertine to striking effect. A large *cabreo*, or estate map, of La Foce, dated 1837, still hangs on the wall and a giant, twenty-six-sided stone polyhedron is set at the foot of a gently curving travertine staircase leading to the upper floor. (Generations of children have slid down the travertine parapet and attempted to dislodge the heavy ball, which is periodically re-cemented to keep it in place.) On the wall of the staircase, visible only to those actually on the stairs or looking down from the upper landing, is an estate map drawn by Pinsent. As the years went by (until 1941), he modified it to show all the improvements to the estate, including the new farmhouses, the schools, the clinic, and the cemetery—a precious chronicle of La Foce history under Origo stewardship.

In 1929, the dining room was adorned with delicate imaginary trompe-l'oeil landscapes framed by loggia-like arches and a Venetian-style balustrade above, where the doors are painted with shells and scrolls, and flowery garlands in swirls of cream, gold, and pale blue lead to attics, cupboards, and stairways. Pinsent's drawings were carried out by a local painter, who declared himself able to paint anything—except the leaves of trees, or so the story goes. We use this dining room for formal occasions and family celebrations—at Christmastime a fire roars in the vast travertine fireplace surmounted by the Origo family crest and draped with Della Robbia–style garlands of oranges and lemons.

At the very beginning of their life at La Foce, Antonio and Iris created a Christmas tradition that embraced the *mezzadri* and their children, who were all invited to a wondrous party on Christmas Eve. The elaborate preparations were planned and supervised by my grandmother, who loved organizing any kind of event, especially if there were children involved. A huge Christmas tree was installed, adorned with English-bought ornaments and blazing candles—a break from Italian tradition and an exotic wonder for the poverty-stricken farmhands of the Val d'Orcia. A Nativity play was rehearsed weeks before by the schoolchildren, special costumes were fashioned by the women who worked in the villa (I remember yearning after a sparkling angel dress), and elaborate stage sets were created by the La Foce carpenter (a papier-mâché ox head has survived and hangs in one of the La Foce houses).

My grandfather contributed to the festivities with a Christmas crib featuring eighteenth-century Neapolitan

---

OPPOSITE *A painted wooden chest in the frescoed dining room.*

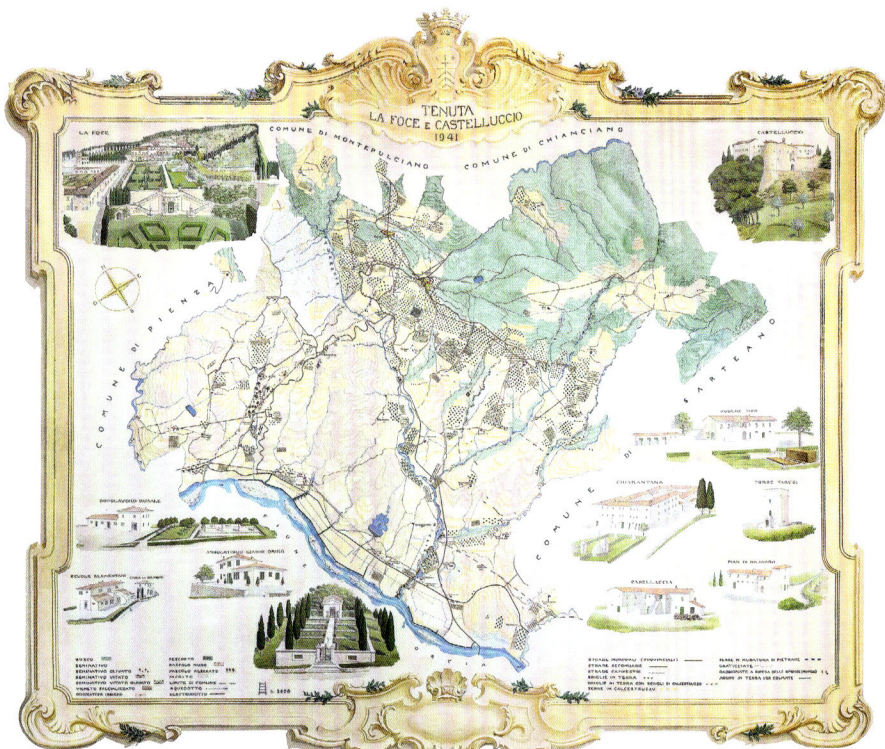

LEFT
The estate map by Cecil Pinsent is painted on the wall above the main staircase and is dated 1941.

OPPOSITE
Detail of an old estate map, or cabreo, *dated 1837, that hangs in the entrance hall.*

wooden statuettes that had long been in the Origo family—one essential preparation for the occasion consisted of an expedition to the woods to find the fluffiest, greenest moss to lay them on. At a certain point in the party, Antonio would disappear to don a red velvet costume and a long white beard. When loud knocking was heard from the balustrade above and Father Christmas made his appearance, dragging a large sack brimming with presents and calling out the names of the children, even the bravest ones hesitated and several burst into tears—especially when required to place a kiss on those rouge-reddened cheeks in order to claim their gift. Local people still tell me stories handed down through the generations about a doll brought all the way from America, with eyes that opened and shut, or a Swiss children's watch with a leather strap. Later, when everyone had gone, the family sat down to a dinner table laid with Christmas crackers from Harrods, a stuffed roast turkey, and a Fortnum & Mason Christmas pudding ringed with blue flames. The tradition continues today—with the exception of the Christmas pudding, which nobody ever liked.

Work on the villa progressed rapidly, and in 1932 a new suite of rooms was added in another wing of the villa. Iris's bedroom and dressing room were strategically placed on the upper floor to provide the best views of the garden and decorated in her favorite shades of blue and green. Pinsent designed the large wardrobe and exquisite matching dressing table painted in sea green. Some of my most cherished childhood memories are of trying on my grandmother's pearls while closed inside the wardrobe, whose folding mirror panels gave the illusion of a secret hiding place and allowed me to admire myself from every possible angle.

At the same time (during 1931 and 1932), Pinsent also created a harmonious complex out of the jumble of small and large buildings adjacent to the villa, used for storage and various farming necessities. Photographs from the early days of my grandparents' ownership show an area—today's *cortile*, with a well at the center—that is almost unrecognizable: La Foce was always the hub for all agricultural activities of the farms on the estate, and the Origos were determined that it

should continue to be so, but at the same time Pinsent was tasked with creating a more pleasing overall appearance than the original buildings Iris described in 1924:

> *Beneath the house stood deep wine-cellars, with enormous vats of seasoned oak, some of them large enough to hold 2,200 gallons, and a wing connected the villa with the fattoria (the house inhabited by the agent or fattore and his assistants) while just beyond stood the building in which the olives were pressed and the oil made and stored, the granaries and laundry-shed and wood-shed and, a little further off, the carpenter's shop, the blacksmith's and the stables. The small dark room which served for a school stood next to our kitchen; the ox-carts which carried the wheat, wine and grapes from the various scattered farms were unloaded in the yard. Thus villa and fattoria formed, according to old Tuscan tradition, a single, closely-connected little world.*[33]

The simple geometric forms of the old farm buildings were dignified by generously proportioned arches and window openings to create a clearly defined and harmonious architectural language and identity, which Pinsent then repeated throughout the entire estate.

The importance of Pinsent's role in creating the "new" La Foce cannot be overstated. Not only did he renovate and extend the villa and the *fattoria*, he also designed the nursery school and a primary school, the small clinic dedicated to the memory of Gianni, the cemetery, and the *dopolavoro*, the social club for the workmen on the estate. He designed a model farmhouse to replace the crumbling, half-ruined buildings, which was replicated many times as the number of farmhouses increased from twenty-nine to fifty-seven over the years. He was called on for advice when the *Consorzio di Bonifica* planned major agricultural projects that would reshape the landscape, as in the case of the iconic cypress-lined road built on the steep reclaimed land opposite La Foce, created to provide the new farmhouses with a direct route down to the *fattoria*. Little more than a stony track that zigzags its way up the hill, it is nevertheless a central element of the view from the wisteria pergola in the upper part of the garden and a tribute to the trecento and quattrocento frescoes of Ambrogio Lorenzetti and Benozzo Gozzoli. As the British landscape historian John Dixon Hunt remarks:

> *Practical exigency has been turned into a striking aesthetic gesture, yet an aestheticism that springs subtly from practical circumstance and tradition. For the road and its tell-tale cypress markers have perhaps another role to play—they tell of a landscape so cared for, so enhanced, that it is nothing short of a garden itself: Tuscan garden as designed landscape. If it recalls the wonderful, preternatural landscape through which the Three Magi travel in Benozzo Gozzoli's frescoes in the chapel of the Medici Palace in Florence . . . this is no accident: La Foce's zigzag relates modern agricultural successes to earlier myths of Tuscany's natural perfections.*[34]

The "zigzag road," as we always call it in the family, was proudly claimed as a symbol of land reclamation at La Foce, and my grandmother would explicitly draw a parallel with the magical landscape depicted in Ambrogio Lorenzetti's *Allegory of Good Government*. At La Foce, the past is never very far away. Some years ago, while looking through the attics I came upon a box of enchanting children's books once belonging to Gianni — among them a volume published in 1929 whose engraved blue cover echoes the title: *The Kingdom of the Winding Road*. On the flyleaf, a note in my grandmother's handwriting: "Gianni, from Granama" (Iris's American grandmother).

OPPOSITE *The opening day of* Dopolavoro La Foce, *May 1939.*
ABOVE, TOP TO BOTTOM *The clinic, dedicated to the memory of Gianni Origo and built in 1933; an old farmhouse; and a new farmhouse designed by Pinsent.*

Alas, Pinsent's drawings, and indeed all the plans of the house, were destroyed during the war along with other family documents. In June 1944, they were taken for safekeeping to Pietraporciana, one of the more remote farmhouses, where Iris and Antonio planned to bring the group of *contadini* and refugee children sheltering at La Foce if the retreating Germans bombed the villa. In a tragic irony, the bombs fell on Pietraporciana, whose strategic position at the top of a hill caused it to be chosen as a base by the local partisans. When it was all over, several partisans as well as members of the Scots Guards and Coldstream Guards had been killed, and all that was left of Pinsent's plans were a few scraps of charred paper, found in a nearby field.

Fortunately, a few blueprints of the garden and some of its principal architectural features have survived—the grand double staircase of the lower garden, the fountain, urn, and baroque bench—underlining Pinsent's fascination with geometry and proportions. ◉

ABOVE *Gianni's American great-grandmother gave him this book in 1929.* RIGHT *The so-called zigzag road was inspired by Sienese and Florentine frescoes of the fourteenth and fifteenth centuries—it has since become an icon of Tuscany.*

THE
HOUSE

PAGE 66
An abundance of ivy geraniums (Pelargonium peltatum) *adorns the La Foce courtyard.*

RIGHT
Aerial view of the villa and part of the gardens.

LEFT
*The travertine steps
leading up to the
front doorway and
entrance hall.*

FOLLOWING PAGES
*The stairway leading
down to the garden
and the view from
the entrance hall.*

LEFT
The old cabreo, dated 1837, hangs on the wall in the entrance hall.

OPPOSITE
Pinsent's huge twenty-six-sided travertine polyhedron is set at the foot of the central staircase.

FOLLOWING PAGES
Bronze sculptures by Clemente Origo, Antonio's father, on the seventeenth-century table in the entrance hall.

PRECEDING
PAGES, LEFT
The entrance hall is paved with terracotta tiles framed by travertine slabs.

PRECEDING
PAGES, RIGHT
View of Pinsent's polyhedron seen from the dining room.

RIGHT
The dining room at La Foce.

OPPOSITE
A detail of Pinsent's trompe-l'oeil frescoes of imaginary landscapes in the dining room.

BELOW
A wooden sideboard in the Renaissance manner, designed by Pinsent.

FOLLOWING PAGES, LEFT
Detail of a painted wooden chest and an imaginary landscape.

FOLLOWING PAGES, RIGHT
Scrolls, garlands, acanthus leaves, and the Origo coat of arms above the imposing travertine fireplace.

RIGHT
A Venetian-style balustrade runs along the upper part of the dining room.

FOLLOWING PAGES, LEFT
Doors painted in swirls of cream, gold, and pale blue lead to attics, cupboards, and stairways.

FOLLOWING PAGES, RIGHT
Scrolls, acanthus leaves, and garlands of fruit and flowers adorn the painted doors.

86

PRECEDING
PAGES, LEFT
The painted doors in the dining room open onto bedrooms, corridors, and more stairways.

PRECEDING
PAGES, RIGHT
A bedroom at La Foce.

OPPOSITE & RIGHT
In Antonio's study, paintings and studies of horses by his father, Clemente Origo, adorn the walls. Clemente was a well-known painter and sculptor as well as a cavalry officer. His portrait by Julian Russell Story hangs at center in the opposite photograph.

PRECEDING
PAGES, LEFT
Clemente Origo's plaster mold for his sculpture of the head of St. John the Baptist.

PRECEDING
PAGES, RIGHT
A view of the wisteria pergola in the fountain garden, with Mount Amiata in the background.

LEFT
The music room at La Foce.

PRECEDING
PAGES, LEFT
*Detail of the eighteenth-
century fortepiano
in the music room.*

PRECEDING
PAGES, RIGHT
*The piano in the
music room.*

RIGHT
*The ground-floor
sitting room.*

OPPOSITE
Travertine and terra-cotta tiles are used throughout the house.

RIGHT
A portrait by an unknown artist of Benedetta and Donata Origo, 1950.

LEFT
Katia Lysy in the library at La Foce.

FOLLOWING PAGES
Details of the library.

LEFT
The travertine staircase in the entrance hall.

OPPOSITE
Cecil Pinsent's map of La Foce was completed in 1941 and shows all the improvements made to the estate.

OPPOSITE
Detail of Pinsent's estate map.

RIGHT
View of the lemon garden from Iris's bedroom on the upper floor.

FOLLOWING PAGES
Iris's sitting room. The collection of Oriental prints, vases, and netsukes came to Iris from her American grandparents, the Cuttings.

PRECEDING PAGES
Details of Iris's bedroom.

LEFT
*Iris's bedroom was
strategically placed
on the upper floor
to provide the best views
of the garden.*

LEFT
Detail of Iris's bedroom.

OPPOSITE
Iris's writing desk.

FOLLOWING PAGES
Iris's dressing room and bathroom were decorated in her favorite shades of green and blue. The dressing table was designed by Pinsent.

LEFT
Iris's bathroom, with blue Vietri tiles.

OPPOSITE
The black marble washbasin was designed by Pinsent.

FOLLOWING PAGES
Arches and pastel colors are dominant elements in the corridors.

PRECEDING PAGES
La Foce bedrooms.

LEFT & OPPOSITE
The painted wooden bed and the English chintz date to the 1920s.

FOLLOWING PAGES
The kitchen at La Foce, in the Tuscan tradition.

LEFT
Pinsent's limonaia *was completed in 1938.*

ABOVE
A dance was held on the lawn in front of the limonaia *to celebrate the baptism of Benedetta Origo in September 1940.*

FOLLOWING PAGES
Interior of the limonaia.

RIGHT
*The Peruzzi-style facade.
La Foce was built in 1498
as a posthouse, or tavern.*

LEFT
The cortile, or courtyard, with a well at its center.

CHAPTER 4

A SETTING OF BEAUTY, A LABOR OF LOVE

Cecil Pinsent created the garden at La Foce in, broadly speaking, four stages from 1926 to 1939. First came the fountain garden, with its cupola-shaped laurel grotto and box hedges—an immediate refuge from the dust and flies and mud, a place for Iris to read and for Gianni to play with his favorite toy, a clockwork frog. The lemon terraces were started in 1933; in 1938 the rose garden above them and the hillsides beyond were landscaped; and in 1939, on the eve of war, the lower garden with its travertine grotto and double staircase was added.

By the time he came to tackle the geographical and climatic difficulties of La Foce, Pinsent was a master of the hybrid Anglo–Italian garden with its terraces, balustrades, and fountains, inner box-hedged *stanze* and wisteria-covered pergolas, all employed in the planning of his Florentine gardens at I Tatti (for Bernard Berenson), Le Balze (for Charles Strong), and Villa Medici (for Iris's mother).

Though Iris longed for a garden, she was initially held back by the Val d'Orcia's chronic lack of water and the need to prioritize agricultural projects. In *Images and Shadows*, she tells us how the irrigation problem was solved:

In the year after our marriage, my American grandmother—somewhat startled to find herself, in mid-summer, in a house in which there was so little water, even in the baby's bathroom—presented us with the wonderful gift of a pipe-line which, leading from a spring in the beech-wood at the top of our hill (some six miles away) brought us our first abundant water-supply.[35]

In 1927, the first part of the garden was completed: a lawn defined by laurel hedges and smaller geometrical box hedges, centered around a stone fountain set atop two dolphins (probably seventeenth-century but of unknown provenance) in a basin with the rococo shape and curbed edge Pinsent favored. My grandmother planted enormous quantities of irises, daffodils, hyacinths, pansies, forget-me-nots, dahlias, zinnias, and chrysanthemums, timed to bloom throughout the seasons, as well as tulips in contrasting colors—deep red and white—inside the box hedges. Now that the box hedges have grown too tall for any but the largest plants to be visible at ground level, my mother, Benedetta Origo, La Foce's present owner, has replaced the tulips with Perovskia sage, whose free, loose structure contrasts pleasingly with the sharp edges of the box. In one corner of the lawn, a laurel grotto provides cool, shady refuge from the summer heat, complete with a travertine seat and table. A pergola of Mermaid roses and wisteria supported by travertine columns acts as a connection between the old part of the house and the new wing. As designed by Pinsent, the travertine-paved area adjoining the house was too narrow

PAGE 140 *In the light of the sunset, the wisteria on the pergola takes on a rose hue.* BELOW *Iris and her son, Gianni, in the fountain garden, completed in 1927.* OPPOSITE *The stone fountain at the center of this garden has two dolphins (probably from the seventeenth century) and a basin with the rococo shape and curbed edge favored by Pinsent.*

RIGHT
Aerial view of the fountain garden, the lemon terraces, and the rose garden, along which runs the wisteria pergola.

for use as a terrace, so my mother extended it into an outdoor dining and seating area.

The years 1929 and 1930 brought other additions and adjustments to the layout of the garden. The old road was moved away from the front of the villa to create an open area marked by pillars with urns and polyhedrons, making room for a gate and a short entry drive lined with cypresses. A forecourt was filled with gravel and planted with ilex trees clipped in an umbrella shape. Before visitors enter the gate, a low wall to the right, topped with the omnipresent twenty-six-sided stone polyhedrons, offers a sweeping view of the estate, with the *crete* still very much in evidence (they are now a protected feature of the landscape, and no farming or grazing of animals is permitted on the hills). When the visitor turns from the larger landscape to enter the formal court, he is presented with the arches and window openings of the Peruzzi-style facade, and only progresses gradually to the garden—more intimate, private and not so readily available—in a passage from formal to informal.

Work on the lemon garden began in 1933. From the fountain garden, two majestic pillars topped with urns mark the entrance to a series of terraced rectangles enclosed by box hedges. These green *stanze* are edged with lavender and columbine at the cut slopes and banks, while pomegranates, hypericum, lilac, and climbers such as honeysuckle, clematis, and roses are planted along the south-facing wall that divides the lemon garden from the fountain garden. In spring, large terra-cotta pots with lemons are brought out and placed on the round stone bases set within the *stanze*, then wheeled back into the *limonaia* (completed by Pinsent in 1938), to help them survive the rigors of the Valdorcian winter. A deep flowerbed planted with peonies, clematis, wisteria, pomegranates, leadwort, and other perennials runs along a steep wall topped with Pinsent's characteristic urns, bordering the far side of the lemon garden and protecting it from the road

OPPOSITE *In spring, the lemon trees in their large terra-cotta pots are brought out from the* limonaia *and placed on round travertine bases.* RIGHT *Work on the lemon garden, a series of terraced rectangles enclosed by box hedges called* stanze, *began in 1933.* FOLLOWING PAGES *The clay hills of the Val d'Orcia are visible beyond the green* stanze *of the lemon garden.*

running below. Around the box-hedged rectangles, the path is paved with the local travertine marble, quarried from nearby Rapolano and used throughout the garden. Originally a dazzling white (as the old photographs of the building of the lower garden and double staircase show), this is a very porous stone that almost immediately fades to a pleasing, uneven shade of gray dotted with moss and lichens. In the cracks between the paving stones, vibrant aubretia, alyssum, campanula, and valerian are allowed to grow freely.

The other side of the lemon garden follows the curve of the wisteria pergola in the upper garden—it is shadier here and the big white Annabelle hydrangea dominates the flowerbed running along the wall. Inside a rectangular lawn, also bordered with box hedges and paved with a travertine pathway, a shady bower of wisteria and banksia roses was one of my grandmother's favorite places to read and entertain guests. The massive stone table and wrought-iron chairs with their art nouveau curves and blue-green hue are the work of Pinsent, as are all the benches in the garden, though in later years my grandmother found them uncomfortable and preferred to recline on a wicker chaise longue.

In 1938 Pinsent created the uppermost level of the garden, with a rose garden and a long pergola (the *voltabotte*, literally "barrel vault," consisting of narrow wooden poles and metal hoops) covered with *Wisteria sinensis*, banksia roses, and grapevines. Rising above the lemon garden, the pergola frames the rose garden and winds its way around the hill as far as the edge of the woods. Halfway along, it opens into a belvedere where, on one of Pinsent's baroque stone benches, my grandparents would sit at the end of the day, admiring the spectacular view of Mount Amiata, the zigzag road, and even the distant tower of Radicofani. Especially in wartime, this was where Iris and Antonio would come to discuss difficult decisions, weighing the risks to their workers living on the outlying farms against the necessity of asking them to shelter escaped Allied prisoners and partisans on the run. During the war, my grandmother would take her baby daughter Benedetta along this path in a pram stuffed with maps, clothes, and shoes to

PRECEDING PAGES *Box hedges border the rectangular* stanze *in the lemon garden.* ABOVE & OPPOSITE *The* voltabotte, *or wisteria pergola, in 1938 and today.*

the nearby woods where partisans, Allied soldiers, and Fascist deserters alike awaited her help.

Above the rose garden, two distinct allées climb the hill. One begins in the lemon terraces and continues beyond the rose garden between rows of cypresses, ascending steeply through steps cut into the soil and marked by travertine risers. The other is both more rustic and more complex, featuring a series of curved stairs (again, only stone risers mark the steps) around elliptical terraces, planted with alternating cypress and stone pine trees. The allées converge on a seventeenth-century statue of a man with a cornucopia filled with fruit and grapes, possibly an allegory of autumn. This statue by Orazio Marinali (1643–1720), placed on a base designed by Pinsent and flanked by two stone benches, is part of a series representing the Seasons. (His companion, a male figure bearing a weighty bundle of agricultural implements and sheaves of wheat on his back, is located at the far end of the lower garden.)

The hillside above the rose garden is cut into a series of terraces:

> *The rest of the hill above has been gradually transformed into a half-wild garden with Japanese fruit trees and Judas-trees, forsythia, philadelphus, pomegranates and single roses, long hedges of lavender and banks fragrant with thyme, mint and absinth, and great clumps of broom.*[36]

Here, the dark red, highly scented roses my grandmother loved (we have been unable to discover their name) contrast beautifully with the blue haze of the lavender and the white Japanese anemones.

In 1939 Pinsent and Iris added another enclosed formal garden, a sunken, sloping terrace with a grotto continuing along the axis from the house through the lemon garden. At the end of the garden, behind an octagonal basin, the focal point is the other statue by Marinali (thought to be an allegory

---

PRECEDING PAGES *A view of the rose garden, now planted also with peonies, irises, salvias, and allium.* LEFT *From the rose garden, an allée flanked with cypresses climbs steeply to a seventeenth-century statue.*

A SETTING OF BEAUTY, A LABOR OF LOVE

of summer) placed on a base that is also an elaborate bench. From here, the view is monochromatic: green grass, green box hedges, and green double walls of cypress and ilex—no other colors except for the white travertine backdrop of stairs and grotto. It is an architectural garden, created by Pinsent in the grand manner of sixteenth- and seventeenth-century Italian gardens. My grandmother's love of flowering plants and color, so evident in the upper garden, is absent here except along the wall lining the balustrade, where Pinsent's urns, this time without their lids, are planted with scarlet geraniums in a last burst of color. When viewed from above, from the paved belvedere at the top of the double staircase, the theatrical effect is striking—as if from the bow of a ship, the gaze embraces a panoramic view of the Val d'Orcia. Antonio and Iris would sometimes come here to dine "on summer nights when, just before the harvest, the whole garden would be alight with fireflies and the air heavy with nicotiana and jasmine."[37]

In an ideal partnership between architecture and landscape, the gardens capture the genius loci and evoke emotions that linger in our memories. One hundred years have passed since La Foce came into the care of Iris and Antonio, and although a garden is a living thing, very little has changed. Some trees have died and been replaced, of course, and a fence has been erected all around the garden as protection against deer, boar, and porcupines (who especially love the iris bulbs). We are ever more fearful of the threats caused by climate change, though the spring that feeds the garden, still collected in a large reservoir in the woods built by my grandfather in the 1950s, has, for now, stood the test of increasingly long, dry summers. We do not cut the grass and wildflowers on the terraces above and around the garden until they dry out and turn yellow, to encourage insects and bees and butterflies. We use organic fertilizers and pesticides—La Foce is an organic agricultural estate. Gazing out on our paradisiacal corner of Tuscany, I can think of no better ending to this book than to quote Iris once again: "I do not think that either of us has seriously wished that we had chosen to live somewhere else, nor to lead another sort of life. The fascination of the Val d'Orcia held—and still holds."[38]

LEFT & OPPOSITE *The lower garden, in 1939 and today.*

THE
GARDEN

PAGE 160
Peonies line the walk along the steep wall bordering the bottom of the lemon garden.

RIGHT
Panoramic aerial view of the La Foce gardens embracing the reclaimed landscape of the Val d'Orcia and its symbolic mountain, Mount Amiata.

PRECEDING PAGES
Two majestic pillars topped with urns mark the boundary between the fountain garden and the lemon terraces.

LEFT
Box cupolas and rectangles in the lemon garden.

OPPOSITE
The lemon garden and a view of the fifteenth-century wing of the villa.

LEFT
Pinsent's urns, without lids, mark the start of the paved belvedere.

OPPOSITE
The lemon garden is divided into box-hedged rectangles or stanze.

FOLLOWING PAGES
A steep wall topped with urns runs along the far side of the lemon garden.

PRECEDING PAGES
Yellow and rust-colored moss and lichen cover Pinsent's characteristic urns.

LEFT
A deep flowerbed planted with peonies, clematis, wisteria, pomegranates, and other perennials.

OPPOSITE
Peonies are planted throughout the La Foce gardens.

PRECEDING PAGES
In 2017, a laurel hedge was cut back to repair the wall. It has since grown back to its original height.

OPPOSITE
Travertine steps lead up to a seventeenth-century statue by Orazio Marinali, probably an allegory of autumn.

RIGHT
Travertine steps lead from the wisteria pergola down to the lemon garden.

LEFT
Peonies and travertine steps in the lemon garden.

RIGHT
Big white "Annabelle" hydrangeas dominate the flowerbed running along the wall. All the benches in the garden were designed by Pinsent.

FOLLOWING PAGES
The flowerbeds in the rose garden form an intricate geometrical pattern and are edged with travertine.

LEFT
The Perovskia, or Russian, sage begins to flower when the lavender starts to fade.

FOLLOWING PAGES
White and pink Japanese anemones grow in the flowerbeds flanking the wisteria pergola.

PRECEDING PAGES
From the wisteria pergola, a magnificent view of the lower garden and the landscape, including the famous zigzag road.

OPPOSITE
At the end of the lower garden, an ornate stone bench is surmounted by a seventeenth-century statue by Orazio Marinali, probably an allegory of summer.

RIGHT
View of the lower garden from the grotto.

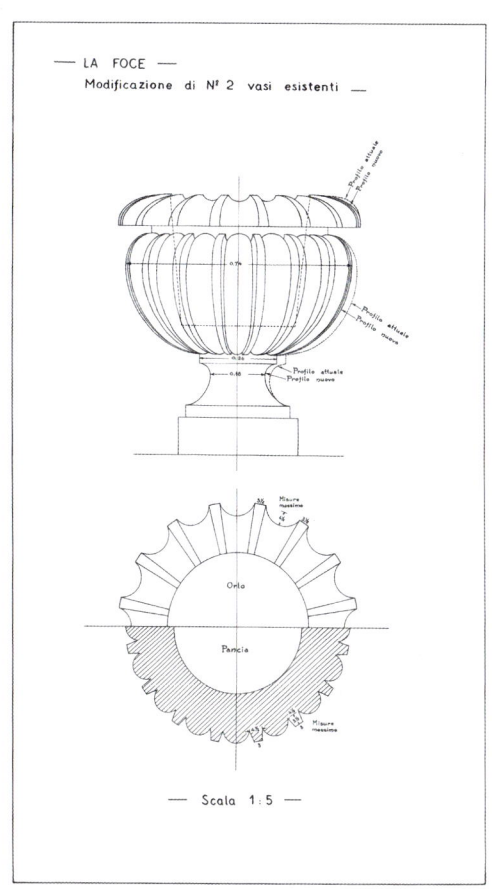

LEFT
Pinsent used travertine from the local quarries at Rapolano throughout the gardens, including for the grand double staircase.

ABOVE
Only a handful of drawings for the garden survived the Second World War. This plan for an urn is one of them.

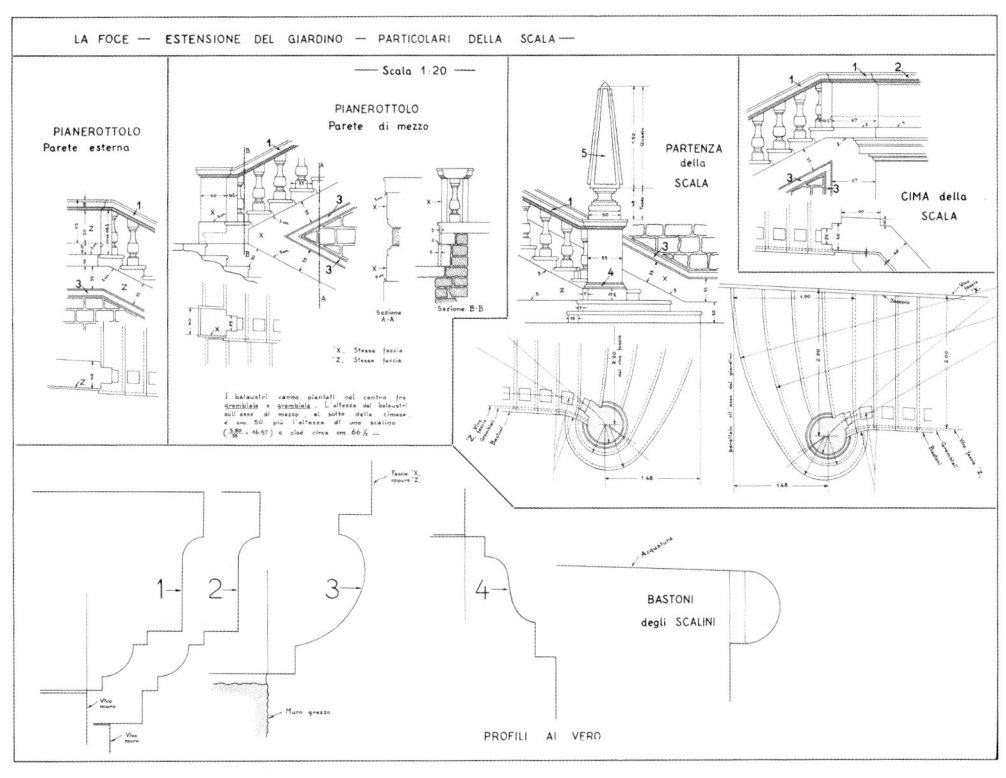

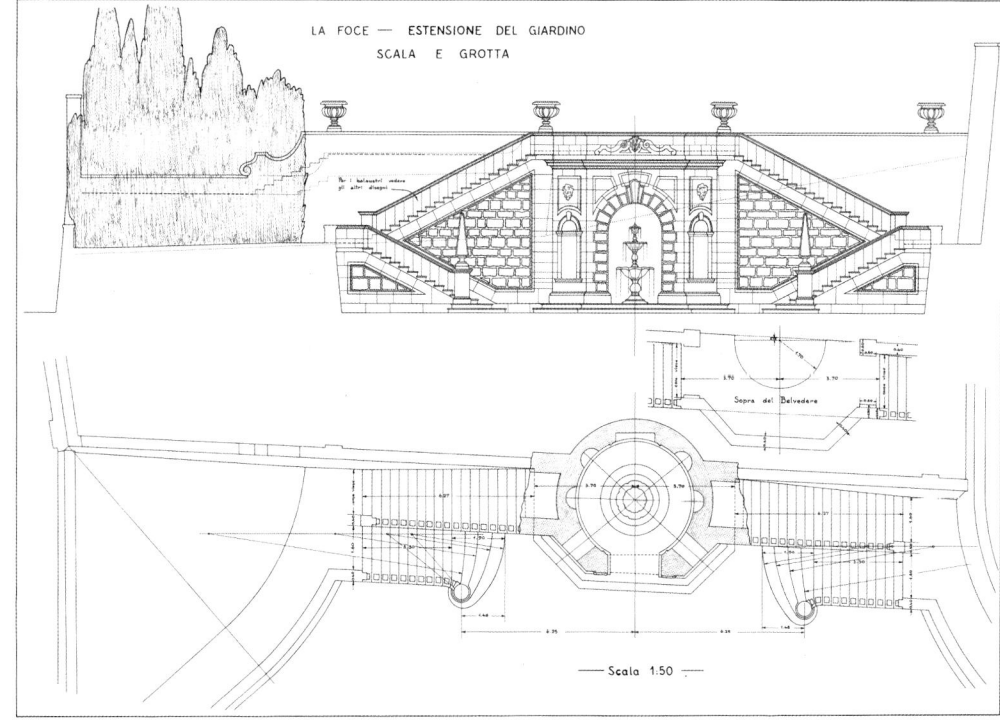

OPPOSITE
The baroque-style fountain in the lower garden.

RIGHT
Pinsent's drawings for elements in the lower garden.

PRECEDING PAGES
Seen from the bottom of the lower garden, the upward slant of the site is evident.

BELOW
Drawings by Pisent of the bench and pedestal for the statue and the fountain in the grotto.

OPPOSITE
The ornamental basin at the end of the lower garden. Orazio Marinali's statue is probably an allegory of summer.

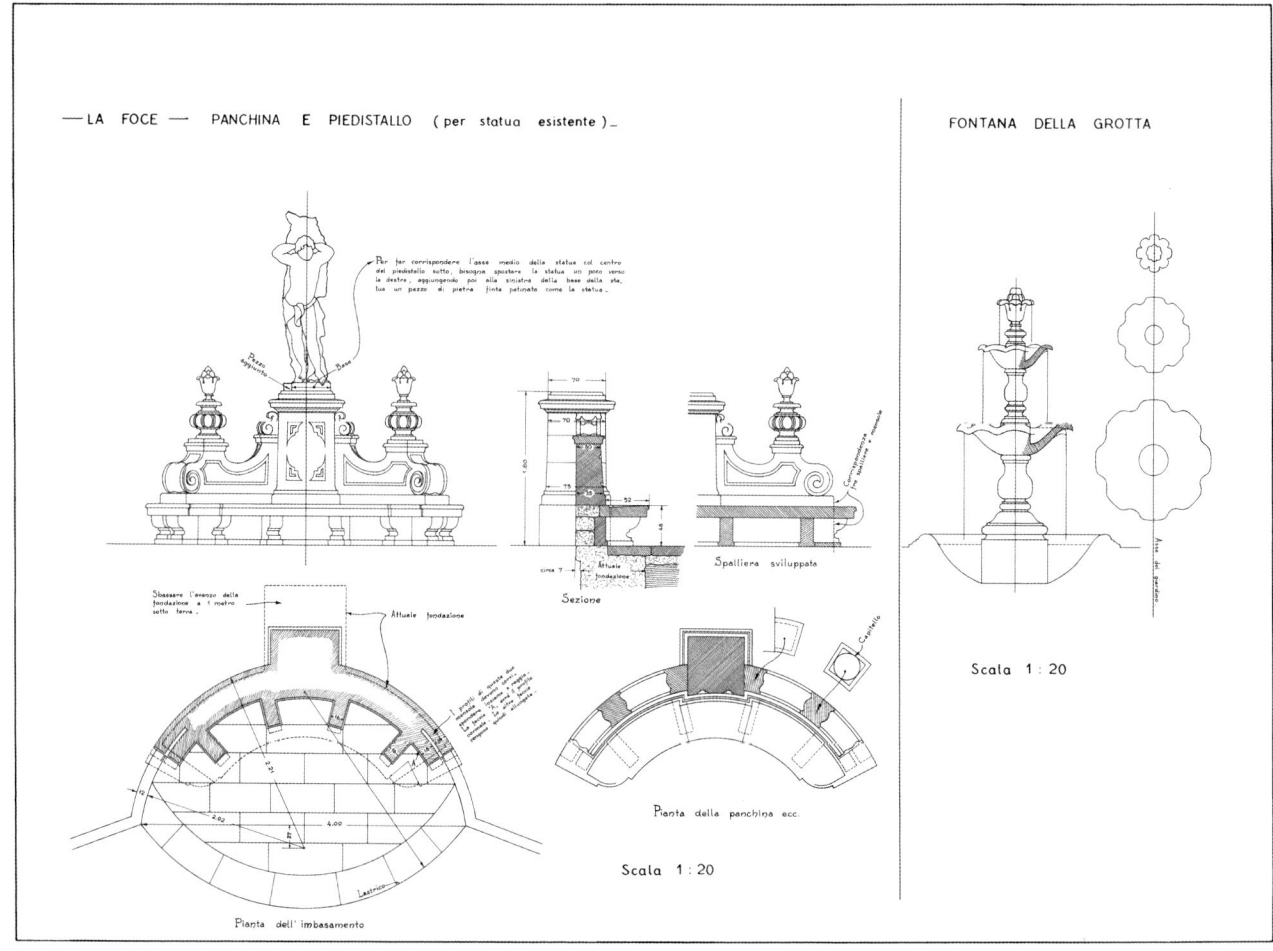

OPPOSITE
From above, the view is monochromatic: green grass and green box hedges—no other color save the white travertine backdrop of stairs and grotto. The dash of color provided by the red geraniums in Pinsent's travertine urns was added much later, in the 1990s, by Benedetta Origo.

RIGHT
Orazio Marinali's statue of a male figure bearing a cornucopia filled with fruit and grapes, probably an allegory of autumn.

ACKNOWLEDGMENTS

FOR THE FIRST PERSON TO THANK I need look no further than my mother, Benedetta Origo, passionate custodian of the La Foce legacy for more than forty years. Under her creative, resourceful, and determined leadership, La Foce has weathered many storms and exponentially expanded the number of its admirers. The landscape, history, and living memory of La Foce and the Val d'Orcia are safe in her hands and in those of her sister, Donata Origo, whom I thank for her encouragement and suggestions. My thanks also go to all the people at La Foce and beyond, who have worked tirelessly to help me put this book together. Some of them are descendants of the original inhabitants; others are new arrivals who have fallen in love with the beauty of the valley.

I must also thank for their supportive, gentle guidance the editorial team at Rizzoli, headed by Philip Reeser. His unflagging admiration for the story of La Foce and its protagonists has made it possible to overcome all obstacles. Charles Miers, Rizzoli's publisher, has been an appreciated advocate of the project, and Sarah Gifford has lent her talents to the book's thoughtful design. Marella Caracciolo Chia kindly made the introduction to the publishing house.

I also thank Simon Upton and Matteo Carassale for their captivating portrayals of every aspect of La Foce and its gardens—their sensitivity and participation shine through in their images.

An enormous thank-you to all those who come to visit La Foce. Whether casual newcomers who have stumbled on its beauty by chance, or loyal admirers who return year after year and introduce their friends and relatives to their favorite places, the Origo family welcomes you all, in the memory of Antonio and Iris.

KATIA LYSY

ENDNOTES

1. Iris Origo, *War in Val d'Orcia: An Italian War Diary, 1943–1944* (New York: New York Review Books, 2017), 158.
2. Iris Origo, *War in Val d'Orcia*, 19.
3. Iris Origo, *Images and Shadows: Part of a Life* (New York: New York Review Books, 2019), 275.
4. Iris Origo, letter to Colin MacKenzie dated October 28, 1923, excerpted by Caroline Moorehead in *Iris Origo: Marchesa of Val d'Orcia* (London: John Murray, 2000), 83.
5. Iris Origo, *Images and Shadows*, 290.
6. Iris Origo, *Images and Shadows*, 291.
7. Antonio Origo, "Verso la bonifica integrale di un'azienda in Val d'Orcia: Risultati di dodici anni di lavoro" (Toward the complete reclamation of an agency in Val d'Orcia: Results of twelve years of work). Extract from *Atti della R. Accademia dei Georgofili* (Proceedings of the Royal Georgofili Academy), XV (Florence: 1937).
8. Iris Origo, *Images and Shadows*, 278.
9. Iris Origo, *Images and Shadows*, 308.
10. Iris Origo, *War in Val d'Orcia*, 22–23.
11. Iris Origo, *War in Val d'Orcia*, 314.
12. Iris Origo, *Images and Shadows*, 11.
13. Iris Origo, *Images and Shadows*, 20.
14. Iris Origo, *Images and Shadows*, 125–26.
15. Iris Origo, *Images and Shadows*, 159.
16. Iris Origo, *Images and Shadows*, 187.
17. Iris Origo, *Images and Shadows*, 56.
18. Iris Origo, *Images and Shadows*, 58.
19. Bayard Cutting, letter to Iris Origo dated December 12, 1911, excerpted by Caroline Moorehead in *Iris Origo*, 31.
20. Iris Origo, *Images and Shadows*, 86.
21. Iris Origo, *Images and Shadows*, 88.
22. Iris Origo, *Images and Shadows*, 92.
23. Caroline Moorehead, *Iris Origo*, 92.
24. Iris Origo, *War in Val d'Orcia*, 237.
25. Iris Origo, excerpted by Caroline Moorehead in *Iris Origo*, 84.
26. Iris Origo, letter to Irene Lawley from 1933, excerpted by Caroline Moorehead in *Iris Origo*, 158.
27. Iris Origo, *Images and Shadows*, 332.
28. Iris Origo, *Gianni* (Florence: privately printed, 1933), 64.
29. Iris Origo, *Images and Shadows*, 235.
30. Iris Origo, *Images and Shadows*, 271.
31. Iris Origo, *Images and Shadows*, 271–72.
32. Iris Origo, *Images and Shadows*, 277.
33. Iris Origo, *Images and Shadows*, 277.
34. Benedetta Origo, Morna Livingston, Laurie Olin, and John Dixon Hunt, *La Foce: A Garden and Landscape in Tuscany* (Philadelphia: University of Pennsylvania Press, 2001), 286–87.
35. Iris Origo, *Images and Shadows*, 347–48.
36. Iris Origo, *Images and Shadows*, 348.
37. Iris Origo, *Images and Shadows*, 348.
38. Iris Origo, *Images and Shadows*, 350.

BIBLIOGRAPHY

Clarke, Ethne. "A Biography of Cecil Ross Pinsent, 1884–1963." *Garden History* 26, no. 2 (Winter 1998): 176–91. Gardens Trust, 1998.

———. *An Infinity of Graces: An English Architect in the Italian Landscape*. New York: W. W. Norton, 2013.

Fantoni, Marcello, Heidi Flores, and John Pfordresher, eds. *Cecil Pinsent and His Gardens in Tuscany*. Florence: EDIFIR Edizioni Firenze, 1996.

Mammana, Antonio. *La scuola rurale della Foce*. Pienza, 2021.

Mazzini, Donata, and Simone Martini. *Villa Medici a Fiesole: Leon Battista Alberti e il Prototipo di Villa Rinascimentale*. Florence: Centro Di, 2004.

Moorehead, Caroline. *Iris Origo: Marchesa of Val d'Orcia*. London: John Murray, 2000.

Origo, Antonio. "Verso la bonifica integrale di un'azienda in Val d'Orcia: Risultati di dodici anni di lavoro," extract from *Atti della R. Accademia dei Georgofili*, XV. Florence, 1937.

Origo, Benedetta, Morna Livingston, Laurie Olin, and John Dixon Hunt. *La Foce: A Garden and Landscape in Tuscany*. Philadelphia: University of Pennsylvania Press, 2001.

Origo, Iris. *A Chill in the Air: An Italian War Diary, 1939–1940*. New York: New York Review Books, 2017.

———. *Gianni*. Self-published. Florence, 1933.

———. *Images and Shadows: Part of a Life*. New York: New York Review Books, 2019.

———. *War in Val d'Orcia: An Italian War Diary, 1943–1944*. New York: New York Review Books Classics, 2017.

Panajia, Alessandro. *Fiesole: esilio di bellezza. Stranieri a Fiesole nei primi anni del '900*. Pisa: Edizioni ETS, 2014.

Pinsent, Cecil. "Giardini Moderni all'Italiana." *Il Giardino Fiorito*, no. 5 (June 1931). Società Italiana Amici dei Fiori, 1931.

Scott, Geoffrey. *The Architecture of Humanism: A Study in the History of Taste*. London: Constable and Company, 1914.

First published in the United States of America in 2024 by
Rizzoli International Publications, Inc.
49 West 27th Street
New York, New York 10001
rizzoliusa.com

Publisher CHARLES MIERS
Senior Editor PHILIP REESER
Production Manager ALYN EVANS
Copy Editor CLAUDIA BAUER
Managing Editor LYNN SCRABIS

Designer SARAH GIFFORD

Copyright © 2024 by Il Valico Gestioni Srl
Text by Katia Lysy
Prologue by Benedetta Origo

All rights reserved. No part of this publication may be reproduced, stored in a retrieval system, or transmitted in any form or by any means, including photocopying, recording, or other electronic or mechanic methods, without the prior written permission of the publisher.

ISBN: 978-0-8478-3624-6
Library of Congress Control Number: 2024931466

Visit us online:
Instagram.com/RizzoliBooks
Facebook.com/RizzoliNewYork
Youtube.com/user/RizzoliNY

The authorized representative in the EU for product safety and compliance is Mondadori Libri S.p.A.,
 via Gian Battista Vico 42, 20123 Milan, Italy, www.mondadori.it

2026 2027 2028 2029 / 10 9 8 7 6 5 4 3 2

PRINTED IN ITALY

PAGE 2 *A spring sunset at La Foce.* PAGES 4–5 *The countryside around La Foce as it is today, with Mount Amiata in the background.* PAGE 7 *Pinsent's travertine pillars topped with urns mark the entrance to the lemon garden.*

PAGE 204 *The white "Annabelle" hydrangeas stand out against the green of box and laurel.* PAGE 207 *A typical summer view of the Val d'Orcia landscape.* PAGE 208 *The so-called zigzag road at La Foce has become an icon of the Tuscan landscape.*

IMAGE CREDITS

MATTEO CARASSALE
2, 4–5, 7, 8, 15–16, 17–18, 24, 54, 64–65, 66, 68–69, 95, 132–33, 136–37, 138–39, 140, 143, 144–45, 146, 148–49, 150–51, 153, 154–55, 156–57, 159, 160, 162–63, 164–65, 166, 167, 168, 169, 170–71, 172–73, 174, 175, 176–77, 178, 179, 180–81, 182–83, 184–85, 186–87, 188–89, 190–91, 192, 193, 196, 198–99, 201, 202, 203, 204, 207

SIMON UPTON
12, 34, 50–51, 58, 61, 70–71, 72, 73, 74, 75, 76, 77, 78, 79, 80–81, 82, 83, 84, 85, 86–87, 88, 89, 90, 91, 92, 93, 94, 96–97, 98, 99, 100–1, 102, 103, 104–5, 106, 107, 108, 109, 110, 111, 112, 113, 114, 115, 116–17, 118, 119, 120, 121, 122, 123, 124, 125, 126, 127, 128, 129, 130, 131, 134–35, 194–95, 208

LA FOCE ARCHIVES
11, 19, 20–21, 22–23, 26, 27, 28, 29, 30, 31, 32, 33, 36, 37, 38, 39, 40, 41, 42–43, 44, 45, 46, 47, 48, 49, 52, 53, 56, 57, 60, 62, 63, 64, 133, 142, 147, 152, 158, 195, 197, 200